ANIMALS IN THEIR PLACES

ANIMALS IN THEIR PLACES

Tales from the Natural World

❦

ROGER A. CARAS

SIERRA CLUB BOOKS
SAN FRANCISCO

The Sierra Club, founded in 1892 by John Muir, has devoted itself to the study and protection of the earth's scenic and ecological resources—mountains, wetlands, woodlands, wild shores and rivers, deserts and plains. The publishing program of the Sierra Club offers books to the public as a nonprofit educational service in the hope that they may enlarge the public's understanding of the Club's basic concerns. The point of view expressed in each book, however, does not necessarily represent that of the Club. The Sierra Club has some sixty chapters coast to coast, in Canada, Hawaii, and Alaska. For information about how you may participate in its programs to preserve wilderness and the quality of life, please address inquiries to Sierra Club, 730 Polk Street, San Francisco, CA 94109.

LIBRARY OF CONGRESS CATALOGING-IN-PUBLICATION DATA

Caras, Roger A.
Animals in their places.
1. Animals.   I. Title.
QL50.C2724 1987   591   86–26019
ISBN 0–87156–707–5

Cover jacket design by Paul Gamarillo
Book design by Wilsted & Taylor
Composition by Wilsted & Taylor
Printed in the United States of America
10 9 8 7 6 5 4 3 2 1

FOR SHEILA AND JOE
*who have added exciting
new dimensions to the clan*

Contents

Acknowledgments

The author would like to thank the following publishers for permission to reprint material included in this book.

The Custer Wolf. Copyright © 1966 by Roger A. Caras. Reprinted with permission of Holt, Rinehart and Winston.

Sarang: The Story of a Bengal Tiger and of Two Children in Search of a Miracle. Copyright © 1968 by Ivan Tors Films, Inc. Reprinted with permission of Little, Brown and Company.

Monarch of Deadman Bay: The Life and Death of a Kodiak Bear. Copyright © 1969 by Roger A. Caras. Reprinted with permission of Little, Brown and Company.

Panther! Copyright © 1969 by Roger A. Caras. Reprinted with permission of Little, Brown and Company.

Source of the Thunder: The Biography of a California Condor. Copyright © 1970 by Roger A. Caras. Reprinted with permission of Little, Brown and Company.

Sockeye: The Life of a Pacific Salmon. Copyright © 1975

ANIMALS IN THEIR PLACES

Introduction

The first task of nature writers is to take their "characters"—the animals, plants, and geological formations about which they write—and put them in their place. Although zoos, heaven only knows, have their many legitimate roles in the conservation scheme of our time, most natural history writers don't write about zoos. Our subjects have not been extracted from their settings by trapping or captive breeding. To be understood, our subjects have to be interactive. An organism or system that cannot interact successfully with its fellow travelers in time and space is known by one awesome, chilling word—*extinct*. Every living thing—speaking of species, here—has to be perfect. Imperfection and survival are mutually exclusive concepts.

If a nature writer or naturalist (we are known by many names) cannot find a creature's place in a system or a smaller system's place in a larger plan, he (or she, of course) cannot in all likelihood put even a single member of the target species in perspective for himself much less his readers.

One of the books used in this collection is called *Monarch of Deadman Bay*. It is about the giant Kodiak bear of the Kodiak Archipelago just south of the Alaskan Peninsula. At the time I undertook the project I was heavily involved in a number of things, all of which seemed to be clamoring louder every day for my undivided attention. I was working on another book that was to be made into a film and Paramount Pictures was expecting me to leave for India, Pakistan (both East and West, in those days), and Ceylon (now Sri Lanka). I also had television commitments, a radio series, some essential meetings in Europe— it was just too much. I had never written about a place I had not been nor at any length about an animal I hadn't seen in the wild and I really didn't want to set a new precedent. But something had to give and it seemed to be *Monarch*. I would write it without ever seeing a Kodiak bear in the wild, without ever even seeing Kodiak Island. I covered the walls of my writing retreat with photographs of not only great brown bears but with drawings and photographs of every species of plant and animal native to the area that I could find. I had maps, tide tables, aerial photographs, and even some botanical specimens on hand. I had searched the literature and had, in skeleton form, the known biology of the great bear. The material I had assembled and pinned and piled around me would give me the setting. I felt certain it would work out well.

I finished *Monarch of Deadman Bay*, version number one, then threw the whole silly thing away—a whole book manuscript—and left for Alaska. After stalking twenty-two Kodiak bears in the wild with my cameras and notepad I was ready to try again. I leave it for others to decide if I succeeded, but I can testify as no one else can that there exists a mighty gap between the frustration of version one and the satisfied feeling that I get from version two. A brief aside here. The Kodiak bear is the only animal I have ever stalked with a gun at hand. The Fish and Wildlife people did not feel I should be in Kodiak bear country, even though in company with a masterful guide, totally unarmed myself. We compromised; I carried a .357 Magnum in a shoulder holster and ignored it all the time I was on Kodiak Island. I do my shooting on pistol and rifle ranges, not in the presence of wildlife.

Why was there such a difference between versions one and two of *Monarch*? There is an enormous difference, obviously, between throwing a manuscript away in disgust and submitting one for publication. The difference lies in what the French call ambiance but I prefer to call texture. For the writer, hopefully, every place has its own sounds and smells, its own feeling, well, its own texture. To write about a Kodiak bear or a lion in the Mara Maasai region of Kenya, you don't have to sit on a rock with a stop watch and record the length of time between matings. That kind of thing is in the literature because of the incredible patience and skill of field biologists. Writers can look up gestation period, average litter size, eye color variations, seasonal dietary and weight changes, and predator/prey relationships. All of that can be had, but no scientific book has "Texture" or "Ambiance" listed in the index. That nature writers have to provide in their endless games of ultimate show-and-tell. Writers have to go to where the animals are and discover or rediscover the place where these animals are interlocked in their perfectly balanced games of life and death. Then they have to interlock that place with the biology of their subjects. Only then do they have their animals in their places and have a story to tell. If I have done that, successfully or unsuccessfully, is for you to decide.

One last point before you go on to the selected readings. To have an animal in its place, the animal has to exist, that is, to be born. Often in my writings we come upon an animal full blown, actively working at being alive. In other cases, wisdom or whatever it is that guides us through this maze called writing dictates that I go to the very beginning. In this first selection a wolf is born and in the pages that follow these (we will dip into them briefly again before we are done) he grows to become the notorious renegade known to history as the Custer wolf.

I

The Beginning of a Legend

From
The Custer Wolf

As from a long sleep, the little wolf—the white one, last of five to be born in the cave beneath the great tree stump—struggled into consciousness. Exhausted from the pulling and crushing, worn out from the first great battle for survival, the little blind animal felt the air rush into his lungs, felt the strangeness of dry air against his nose and throat. Then he slept.

Not much more than a handful of fur, still soaked with the fluids of birth, the small creature felt the joy of movement, experienced the first ecstasy of freedom. Even as he slept his legs twitched and his head turned from side to side. After a few minutes the cub awakened, his whole being possessed with a desire he had not experienced before. There was a gnawing, a need to be filled, a need to press his muzzle and gums against something. He had to suck.

Instinctively he twisted and turned, trying to use his legs that were now free. Finally, driven by the strange mounting need, he crawled on his flat belly toward the source of heat, felt the fur, nuzzled it, searched through it. Something pressed against him, trying to push him back from the fur and warmth, but he struggled mightily against his litter-mates until he reached his mother's body. His nose found a hard little nipple and greedily he pressed himself to it, took it into his mouth and began working his gums. The rich, warm fluid poured into his mouth and he choked. Fighting back against his brothers and sisters, and fighting the great choking at the same time, he finally cleared his throat and mouth and sucked again. This time the fluid flowed properly, and for the first time in his life he swallowed. The aching in his guts stopped. The little white cub had won the second battle of his life. Here, in the hole beneath the ground, he would survive.

Sometime during that first day the wolf cub had a second great awakening. Strange sensations registered on his growing consciousness. With his nose pressed against the fur on his mother's belly, he received an impression that quickly shaped itself into his first memory. The next time he sought the source of warmth and food, there was no need to press into the fur to know that he was close. For now his nose directed his movements. He had the power of scent.

There were other things in the cave besides the fountain of warmth and milk. Something caressed him, cleaned his fur and made it smooth. This was the good feeling, the feeling that went along with the one that stopped the hurting in his belly. Each time he ate, he slept, and when he awakened he felt new strength in his legs. His muscles were filling out and he began to show a measure of coordination. Each time he tried a movement now it came easier. Soon he could crawl to his mother, over and across his struggling littermates, without difficulty. Hour by hour the strength came, and hour by hour the world formed into a reality.

For twelve days, the wolf cubs lived in a world of total darkness. There was no day, no night, only the need to fill their stomachs and the need to sleep. There was the warmth when they were near their mother, and the cold when they weren't. There was that which felt good, and that which felt bad. The great struggle was to achieve the one and fight off the other; as the hours and days progressed, the fight developed from an unconscious one to a conscious experience. Theirs was a completely elemental world composed of life-essentials.

On the thirteenth day a strange thing happened to the cubs. They kept turning their heads involuntarily toward that part of the cave from whence the cold air came. A strange magnet seemed to draw their attention toward the opening. Then, as their sealed eyes slowly began to open, a new sensation was given to them. Suddenly, quite suddenly, there was a light time and a dark time. At first, there was little to be seen, little that could not be recognized better by sound, touch, taste, and odor. Soon, though, there were things that could be separated from others by sight. When the large wolf—the one that did not offer the milk—returned to the cave, when his great form filled the mouth of the tunnel, it could be known because the light was cut off. Indeed, the first visual knowledge the cubs had was of the coming and going of their parents.

Now the world was very real. It could be tasted (by the end of the first week any foot or tail that came close was gummed and mouthed endlessly); it could be smelled; it could be heard. And now, it could be seen.

For four long weeks the cubs were restricted to their

tight little world beneath the tree stump. Smells and even some muted sounds drifted toward them from the mouth of the tunnel as their parents moved in and out of the blinding light, but any move in that direction by a cub brought a warning growl from one or both parents. Those cubs that ignored the growl the first few times quickly learned that a law existed that governed their lives.

One morning both parents left the cave together. The cubs could hear them outside whining, making little calling sounds. Hesitantly they stumbled and rolled over each other toward the sound. Suddenly the father loomed in the light, walked over them, bowling their round bodies against the sides of the tunnel. He turned in the end chamber and began pushing from behind. One against the other, the cubs were tumbled and pushed until they were at the mouth of the tunnel. One by one they felt the searing pain as the hot light crashed against their eyes and blinded them.

Strange odors, strange sounds unheard before, and hot light bombarded them. They whined and turned to re-enter the safety of the tunnel—all but one. The white cub stood alone at the entrance, his forelegs stiff, braced against the world. A baby cry, an approximation of a growl, filled his throat.

Meanwhile, the parents picked up the other cubs in their jaws and gently carried them to the flat place in front of the tunnel mouth. Individually they were retrieved from the tunnel to which they tried desperately to return and were carried again to the world outside. The white cub stood braced until he too was scooped up and carried to the flat place. Tumbled together, cowering as the world exploded its new sensations against them, the cubs huddled and whimpered, each trying to hide under the others. The suspicion and fear with which they reacted to the new world were infectious, and the small, round bundle of white fur that was Lobo whimpered with the rest.

After the first fearful minutes, a new sensation, a new need filled the cubs. They were curious. Despite their fears, despite the threat of the bird sounds in the trees and the wind sounds, they began to investigate. Hesitantly, each turned from the group and began to seek the secrets of this strange new world. Each new sound electrified them and sent them tumbling toward each other. Slowly they became oriented and their curiosity, a passion that would be with them for as long as each lived, took over. Proudly, the parents stood off to the side and watched the awakening.

Within a matter of hours the cubs were rolling in their tumbling games on the flat place. New sounds continued to distract them, but the terror was gone. This was the world to which they belonged. Once the white cub—Lobo—wandered off from the rest and heard the warning growl. Ignoring it, he continued toward the new smell that lured him. Suddenly his father loomed over him and the great mouth closed on his neck. He was jerked off the ground and roughly dropped back among his littermates. This was not the gentle carrying, but a stern punishment for not heeding the warning growl. In the few days that followed, the punishment became more severe, the teeth bit harder into the scruff, and the cubs learned that the warning growl was absolute.

Each day a signal—half whimper, half soft, rumbling growl—told the cubs that they might go to the flat place. And each day a quick growl told them to return to the tunnel. Outside they were allowed to investigate, to move off in new directions for a brief time; the warning growl told them when to stop and turn back, when to lie down, when to follow. Day after day the cubs learned to respond more quickly, came to know more signals, more directing sounds.

There were good times with the grown wolves as well. The tan male would lie down among them in the warm sun and invite them to torment him. They chewed and they snarled, they crawled on him and bit his ears. Patiently the male wolf tolerated their antics; but when he gave his warning signals, compliance must be immediate.

The mother, too, demanded total obedience. Her teeth were sharp, and the press of her great paw was painful on the neck. Together, the parents completed lesson number one in the education of the cubs.

Often the father moved off and was gone for hours at a time. When he returned he would deposit great smelly chunks of meat at his mate's feet. With a satisfied sound in her throat, the female would lie down and gnaw at her food while the male went off with the cubs for the tumbling and fighting games. One day (the cubs' thirty-fourth day) the father brought a small strip of meat to where the cubs were playing and dragged it before their noses. The white cub grabbed at it and began to shake his head, flopping the piece of meat back and forth. The little black female grabbed the other end and began to pull. A mighty tug-of-war was mounted and there was much baby snarling. The harder the white cub pulled, the deeper he sank his needle-sharp baby teeth into the meat.

Suddenly a juice flowed into his mouth and he let go of the prize, sending his sister tumbling end over end. He walked over to where she had dropped the meat and pawed it. He whined and cocked his head from side to side, sniffed it, and ran over toward his father, who was basking in the sun watching the antics. Arching his back, the white cub again stalked the prize; stiff-legged he jumped and rolled over it. Again he bit at the meat. The juice was strange and he ran to his mother and began to suck until the milk had washed the strange taste from his mouth.

Each day their father brought the cubs strips of meat to play with, and each day the games with them grew fiercer. There was the tugging, the pulling and the snarling. The meat could be charged, it could be run with, it could be torn from the grasp of another. But always there was the juice, the taste that came full and rich to the tongue. After a few days the taste was no longer strange. One by one, the cubs learned to gnaw on the prize once it was won from the others. Each time the supply of meat was bigger until there was enough for all.

After a few more weeks—about ten from the time of the birth—the cubs were eating a little meat each day. They still relied to a large measure on their mother's milk, but it was no longer enough. On some days great chunks of raw meat were carried to them by their father within his stomach; arching before them, he would retch the meat up, still fresh, still rich with the juice. On other days there would be a whole animal—a prairie dog or a squirrel—still in its fur. But before such prizes could be ripped apart and eaten, there must be the games; for the meat with the fur still on offered the greatest opportunity for the fighting play.

Always the eyes, the great yellow eyes of the parents, were on them. Always they were watched and often they heard the sound signals, the growls and low throat rumbles, to which they responded. Everything they did had a lesson behind it. In their eating, in their playing, in their very existence, the lesson was always there. One day, while they all lay about the flat place sleeping in the sun, the father leaped abruptly to his feet. The hairs along his back and neck stiffened and his back arched. Motionless he stood with his ears erect and his front legs braced. Suddenly he whirled and growled deeply—the warning sound. There was no allowance in the sound this time, no permissiveness. The female shot toward the mouth of the tunnel and echoed the warning. The cubs stumbled toward her, frolicking over and under each other in mock play. A second time the growl came and the father bit hard on the flank of the small tan male who was slower than the rest. The parent wolves rushed the cubs down the tunnel, herding them into the far chamber. The bodies of the adults blocked the passage, the female on the inside, the male pressed flat near the tunnel entrance, motionless, waiting.

A strange sound began to fill the world. A great thumping resounded through the tunnel and small bits of earth, the dust of the dried walls, filtered down. One of the cubs began to whine and a harsh grunt from the mother cut him short. Terrified, the cubs huddled in the chamber until the great rhythmic pounding faded off. The white cub felt the strange excitement of danger, felt the electricity that ran through his parents; he tried to crawl forward to where his father lay frozen, but a snarl from his mother sent him back to the others.

For the rest of the day the cubs were kept in the tunnel. Their movement was severely restricted and any effort to begin playing was cut off with a growl. For hours they lay in the chamber, until their mother finally came back and stretched out, allowing them to feed. They dropped off to sleep after filling themselves, the great adventure of the day soon forgotten.

The white wolf, the one we call Lobo, did not know that he had had his first encounter with his one and eternal enemy: man.

❧ 2

Breaking Wild Elephants

From Sarang

Fiction is as legitimate a format as nonfiction for telling the stories of animals in their places as long as the writer is willing to travel almost literally to the ends of the earth (a good, hard-working cliché that one should not be ashamed to use when it is applicable) and is able to keep the biology straight.

In a novel as much as in any other type of writing, the story-teller should get his or her readers into the setting as carefully as possible. In the case of Sarang, *the setting was the Chittagong Hill Tracts of what is now Bangladesh, up near the Assam border. Not many people who haven't read Kipling carefully (something of a rarity today) are likely to know that setting or very much about the lore of elephants and how they are converted from wild animals into relatively docile working animals that will spend the remainder of their lives in the company of humans. Sarang is the story of a tiger and two wonderful kids, but the background for their story is the jungle and working elephants. In a sense this was a different kind of exercise. The animals I had to get into place were the readers. I went to the Hill Tracts, spent time with the elephant catchers and trainers, and watched them break new mounts for their lifetime of service in the logging industry. Having gone and seen, I returned to show-and-tell and tried to make the mysterious background of working elephants just a little less mysterious. Then it was time to talk about two kids and a tiger.*

Glenn sat cross-legged on a mat next to Ata eating rice from a small bowl and picking hard-boiled eggs out of a central pot. He sat back and examined the scene around him. Above them the evening wind was soughing quietly through the trees. Somewhere in the distance a coarse, dry cough intimidated a night bird that was whispering a plaintive *seep seep seepadoo.*

"*Chita baghh,*" Satwyne said. "*Hae,*" the others murmured, "*chita baghh.*"

Glenn thought of the leopard out there moving through the forest and felt a slight shiver run through him. He hunched his shoulders against it, and then found himself listening very, very carefully.

You cannot sit quietly in a wild place without becoming a part of the place yourself. A jungle at night particularly absorbs the observer, for it is a whisperer of secrets, a community of powers and mysteries. It is also a place where an elf can become a giant by the simple act of peeling back a shadow. For Glenn it was one of the most exciting and strangely pleasurable experiences of his life. He was out where the animals were, with men whose skill, intelligence and ancient traditions made them masters over the largest terrestrial animals left on this planet.

No one did much sleeping that night, least of all Glenn. At dawn the camp was awake. They ate a simple breakfast of tea and rice in silence while they waited for the first runners to arrive. Glenn looked around at the group sitting quietly by the small fire pit. He couldn't help reflecting how strange it was that some men in the course of their lives come to treat extreme danger with indifference. With the exception of a few score men, anyone faced with the task these men were about to tackle would pale with terror. It is easier to become a United States senator, Glenn thought, than a *phandi*, particularly a *bor phandi*, or top expert, like Satwyne and Oroon, yet their names are in no books, and no newspapers report their doings. Today they will live or die on the job, and few people outside their families will ever know the difference. They are among the most courageous and skilled men in the world, yet there probably aren't a hundred people outside of the Hill Tracts who even know their profession by name. Some eccentric British lord sails his sailboat across an ocean with half the American Coast Guard tracking him on radar and the British Navy out in force to overfly him and the papers are full of it for weeks.

The first runner arrived before they had finished breakfast. A second arrived ten minutes later. The herd was less than a mile and a half away. The elephants were feeding in their direction.

The two *koonkis* were ready to go. Their saddle pads were strapped in place and the jute ropes with the nooses the *phandis* would use were coiled and ready. Each *koonki* was to carry two men into action. The third man on each team would remain behind, his work over for the moment. They were the grass-cutters or *kamlas*, the bottom-rung men. It was their job to supply the *koonkis* with fodder and later to perform the same service for the captives.

In a reversal of the method used in normal elephant operations, the *mahouts*, now the second men on the teams, sat to the rear of the pads holding *ankuses*, long goads with which to guide the mounts. The *phandis*, perched up forward, would be too busy to direct the elephants' movements. Those directions had to come from someone else.

The two teams moved out without ceremony. It was starkly casual. With the exception of an occasional hand signal they would hardly communicate until the first noose was thrown. By some unexplained means, however—and Glenn had heard it described as extrasensory—they would each know what the other was doing and what help was needed. Talking was out of the question. A human voice, even a whisper, is an unnatural sound in the jungle, and a herd with cows and calves moving through dense forest that houses both tigers and leopards is alert to any foreign sound.

Within fifteen minutes the teams made contact. They could hear the herd feeding and from their vantage points high up on the *koonkis'* backs the *phandis* and *mahouts* could see an occasional trunk reach up and search for a special morsel. Also there were the more personal sounds the wild elephants made. The gaseous stomach rumblings known as *borborygmus* were clear and unmistakable.

Satwyne's *mahout*, Ma, automatically held Bolo Bahadur back. As a male he was likely to cause more suspicion than the female Sabal Kali. He was even likely to be attacked by one of the herd's bulls. The cow, however, would be able to insinuate herself and her riders into the middle of the herd without disturbing the routine of the highly suspicious and often dangerously tense animals.

As Oroon's *mahout*, Rauton, eased Sabal Kali in toward

the herd he spotted a young tusker pacing back and forth in a suspicious manner. With almost imperceptible signals he turned Sabal Kali away to approach the herd from another angle. Again, though, he encountered the tusker, who somehow seemed alerted to their intent, and again he had to turn Sabal Kali aside. The tusker was snatching up bunches of debris from the forest floor and throwing them about. He was rocking back and forth on his great forelegs, snapping his ears forward and back and exhaling loudly through his trunk. He took several steps to the right, and then several to the left. He snapped his ears again and blasted air through his trunk, creating a small dust storm at his feet. He took two steps forward, spun around and rushed off into the midst of the herd. Both Oroon and Rauton knew he would be back and would bear careful watching. No one could predict when his clownish display would convert itself into a murderous charge.

Then Oroon saw the cow. She was no more than six feet six at the shoulder but she was unmistakably a *koomeriah*. This cow of the highest caste is one of the best of all working elephants, strong, noble and intelligent. She was their target. She was a young animal, probably one that hadn't yet calved, but she was beautiful. The task that would now occupy all of Rauton's time was getting Sabal Kali close enough to the cow for Oroon to be able to noose her. Oroon hadn't said anything to Rauton, hadn't even signaled to him, but the *mahout* had seen the *koomeriah* at the same time as the *phandi* and had known immediately that as long as that cow was available for a try no other elephant would be of interest.

The cow, unsuspicious, fed calmly at the edge of the herd. She seemed to be particularly attached to an older animal who could very well have been her mother. The two were never very far apart, and both Oroon and Rauton knew that they would have to work their way in between them in order to cut the younger cow out of the herd.

Several more times they were forced to separate from the herd because of the attentions of the nervous young bull. He repeated his threat display twice more and the *phandi* was beginning to find him tiresome.

Although they had seen the *koomeriah* cow before seven in the morning it was almost two o'clock in the afternoon before they finally got close enough to make an approach.

Their patience was rewarded. She stopped to pull some leaves off a high bush and Rauton goaded Sabal Kali forward. Oroon stood on the *koonki's* neck and threw the *phand*. Before the cow knew what had happened it settled over her head and Oroon pulled back hard. The noose was set. The cow immediately spun around and the rope began peeling off the neat pile on Sabal Kali's back. Automatically, without any additional instructions, the *koonki* turned so that the side where the rope was attached to her girth strap was facing the noosed captive. To have allowed the rope to cross her chest or rump would have been to invite a tangle and possible disaster. The *koonki* began sidestepping in toward the bewildered cow who had stopped at the end of the rope. Oroon took in the slack as fast as Sabal Kali presented him with it. The cow stood quietly for a moment trying to understand her predicament, then she took off again. Sabal Kali had to sit back on her haunches to check the flight. Satwyne's *mahout*, Ma, was moving Bolo Bahadur into position between the noosed cow and the rest of the herd. Slowly the two *koonkis* would squeeze her into the position that was necessary to complete the capture.

Satwyne kept his eye on the nervous bull, who was now

thoroughly alerted. The rest of the herd seemed unaware of the situation and were strung off in several directions, feeding contentedly. Ma had to maneuver Bolo Bahadur out of the way twice when the bull seemed intent on charging. An angry tusker could impale and kill a *makhna*. Although the men were anxious to get to Oroon's aid as quickly as possible they had to avoid becoming the target of a charge. If the whole herd stampeded in wild confusion mayhem could ensue. It was a critical moment, with disaster a ready potential. In the meantime Sabal Kali was deftly sidestepping to keep the cow in check and to avoid tangling the rope around a tree. Too much pressure was dangerous for the captive; the noose rested on her trachea. If they allowed her to panic she could easily sustain a fatal injury. A wild lunge against the heavy rope while it was in that position would be the equivalent of a monstrous karate chop across the throat.

Finally, Ma was able to elude the tusker and bring Bolo Bahadur up to the captive on the side opposite Sabal Kali. Quietly and steadily Bolo pushed against the cow and began easing her toward Oroon, who was furiously working to take in the slack and twirl it back onto a neat pile, where it would be free to peel off again without tangling if the captive decided to make another attempt at freedom.

The maneuvering went on for nearly two hours. The tension didn't slacken for a single moment. Both teams had to concentrate on their own strategy, the cow's movements, and the activities of the herd in general. Working in concert they finally managed to separate her from the herd. They could hear the dozen or more other elephants breaking through some brush about a quarter of a mile away. The nervous young tusker trumpeted once. It was a sad sound, a confused lament rather than a challenge or an expression of fury. The men stopped to listen but it was not repeated. The jungle swallowed the herd, now short one beautiful young cow.

The two *koonkis* pinned the *koomeriah* between them, making any further struggle on her part useless. Bracing with their outside legs, they leaned in, putting their combined weight on her flanks. She was already exhausted from her early struggles and went to her knees. She was only down for a moment, though, and was quickly on her feet again. She reached up with her trunk to explore Bola Bahadur's head. Satwyne smacked it smartly and she withdrew it. Her trunk was the weapon the *phandis* feared most for with it she could grasp a man and bring him within reach of her feet. She tried the same trick with Oroon, who also delivered a blow to the sensitive slender end of her trunk. Again she withdrew it.

Satwyne dismounted from Bolo Bahadur and went around to shackle the cow's hind legs while Oroon slipped off Sabal Kali's off side and adjusted the noose around the captive's neck with a check-rope that would keep her from strangling herself. While the two *phandis* were on the ground the *mahouts* urged their mounts to lean in harder, to pin the cow so that she was all but paralyzed. Crushed by the two trained elephants and whacked painfully on the trunk every time she tried to lift it, she was defenseless. She coiled her trunk as tightly as she could and hummed and moaned her despair. Her eyes were wide with terror as she was touched by human hands for the first time.

Satwyne maneuvered the rope around her hind legs, tying a figure eight or *degi* hitch. Later, a similar hitch called a *banda* would be fitted to her forelegs.

With surprising speed the cow was lashed between the two *koonkis*, who were then free to ease the pressure and start slowly walking her toward the *pilkhana*. There her training would commence. Within three weeks she would be expected to carry a *mahout* on her own neck, without a *koonki* in attendance. Her spirit would be broken that quickly, and that quickly she would come to accept the fact that there were masters who were to rule her destiny. She would come to know that pain was associated with misbehavior. She would learn to follow fifteen one-word commands without any display of reluctance. Then, and only then, the brutalizing by man and *koonki* would stop, and for the rest of her life she would be treated not only well but with affection and respect. She was on her way to becoming a trained working elephant.

Once the cow was secured in the *pilkhana* area, Kala and Khada, the two grass-cutters of the *phandi* teams, took over. They brought fodder to her, although it would be a day or two before she would eat, and left tins of water where she could reach them. They poured hot water over the *koonkis*'s backs and rubbed them down, then fed and watered them. They mounted watch on the captive while Satwyne, Ma, Oroon and Rauton returned to the campfire, too exhausted even to discuss the day's activities.

Glenn instinctively knew that these men were too far along in their profession to require a cheering section. They had done what they had set out to do, and had laid their lives on the line in doing it. The fact that they had succeeded and survived was enough.

•   -   •

Glenn's incredible luck held; the two teams were able to cut out and secure a *makhna* before noon on the second day. He was a handsome animal, standing a little over nine feet at the shoulder. While not a true *koomeriah*, he was a fine *dwasala* and appeared to be sensible and even-tempered although apparently immensely powerful. He did not struggle long because the heavy jute rope was cutting his wind off, but twice after the squeeze was started by the two *koonkis* he was able to sweep the powerful Bolo Bahadur aside with a shove of his hip.

The nobility of an elephant is not gauged by the struggle he puts up when captured. In most cases, violent struggling is more a sign of hysteria than of outrage or bravery. The more stable an elephant's nature, the more likely it is that he will hold back to assess the situation before starting to thrash around. What the elephant doesn't understand, of course, is that once his captors are allowed to take the first step his days of freedom are over.

Noosing and securing two fine wild elephants in a little less than thirty hours is a record accomplishment for two *koonki* teams working a *mela shikar* operation in heavy forest—so Ata assured Glenn.

While Kala and Khada attended to the *koonkis* and mounted guard over the captives, the *mahouts* and *phandis* went down to the stream and bathed. Their muscles ached with the tension they had been under since making contact early the previous morning, and they were lathered in sweat. The sweet, cool water flowed over them as they lay almost totally submerged on the sandy bottom, making appreciative noises and calling to each other in cryptic Bengali, laughing at each other's comments about the operation they had just completed. The foolish young tusker who had put on such a show took the brunt of their jokes. They called him *Idur*, the Bengali word for "mouse." The fact that he had put on such a show without attempting to carry through his threats earned him their disdain. He would have been a nervous and unreliable animal in captivity, they assured each other, and it was a good thing that Glennsahib hadn't wanted a tusker. They then went on to discuss the fine qualities of the animals they had taken, particularly those of the beautiful young *koomeriah* cow. Their visitor from Kansas sat on the bank watching them and listening to their conversation. He understood almost nothing of what they were saying, but somehow that didn't matter. It was like being in the locker room with a team that has just won the world series, Glenn thought. He felt highly privileged.

3

A Place for a Giant Bear

From
Monarch of Deadman Bay

It is a toss-up, sometimes, whether to start with the birth of an animal, as I did in the selection from The Custer Wolf, *or to start back a generation as I do here in the selection from* Monarch of Deadman Bay *(version number two!). The decision here was based on the fact that, unlike the Great Plains states (Nebraska and South Dakota) where my wolf story actually did evolve, these, the greatest of all terrestrial carnivores, live on an island and in ways that would be strange to most readers. (That's why it turned out to be so important to go there before telling about it.) A wolf is a much better known animal than a bear, and the plains are far better known than temperate Alaska with its fractured landscapes.*

The decision, then, was to go back to the land, establish the island, start a generation back with the bears, and move forward to the birth of my central character. Then there would be a steady logic to the way the story could unfold with no one, including the author, being out of place or disoriented.

Kodiak Island, all 3,465 square miles of it, huddles in the northern sector of the great Alaskan Bay like an enormous amoeba waiting to envelop the smaller islands of the archipelago that bears this island's name. Cut off from the Arctic Ocean by the Alaskan Peninsula, Kodiak Island is under the comparatively mild spell of temperate southeastern Alaska rather than in the harsher grasp of the land of Eskimos and polar bears.

One hundred and three miles long, fifty-seven miles wide, rain-drenched for much of the year, the Island is situated between 56° 40′ and 58° north latitude, 152° and 155° west longitude. High in the east and covered with conifer and hardwood forests of Sitka spruce and cottonwoods, the Island descends over four thousand feet toward the tundrous west with its uniform cover of muskeg grass. Here only scattered clumps of alders break the monotony.

It is a wild land. Although long settled it has never been tamed. The city of Kodiak, huddling in the northeastern part of the Island, is the sixth largest settlement in Alaska and dates from 1794, when it was the capital of Russian America. Yet only a few miles away it is possible for even an experienced woodsman to get hopelessly lost in a tangle of land that rises and falls like a stormy sea.

This blue-gray land, serenaded by a chorus of gulls a million strong, has over a thousand miles of coastline that resembles, often enough, Norway's fjord-indented shores. The treacherous Shelikof Strait to the north separates the Island from the great land mass of the Alaskan Peninsula by thirty miles. A graveyard for unwary sailors, these waters and those along the Island's other shores roll inward in high, fast tides that swallow the rocky beaches and seaweed flats in greedy gulps. Kodiak Island's granite, slate and even sandstone have so far withstood the sea's intrusion. It owes its shape and the long shadows of its ragged hills to the carving power of prehistoric sheets of glacial ice.

A history measured in millions of years, with the intermingling influences of sea and quake, glacier and wind, makes it hard to tell how a given feature of the land was formed. It may have once been two islands, for the largest fjord to slash her coast, Uyak or Windy Bay, west of the Island's middle in the north, cuts forty miles inland and all but meets Deadman Bay in the south.

Kodiak Island shows many scars, only the least of which were made by man. The quarries that have been cut into her hills, the roads on their sides and the light cosmetic touches to her natural bays are of little account. More momentous things have shaped this land. From June 6 to June 8 in 1912, Mount Novarupta on the Alaskan Peninsula to the north showered millions of tons of raw volcanic ash on Kodiak Island, in many places to a depth of twelve or more inches. All the changes wrought by all the men who have ever stepped ashore on this island are as nothing compared with the force let loose upon it in those few hours.

When the winds are high, and they often are, and when the sea is angry with the land, rain does not simply fall on Kodiak Island; it is hurled against this intruder in the sea like shrapnel. Fog banks engulf her like living things. This island is like a ship at sea, and often she sails uneasily through unaccountable weather fronts. There are days, though, when the land sails out of the mists and drifts on calm and sunlit waters. On such days men know why they have come to this place.

Such a land as this, raw one moment and steaming the next, provides a garden in which giants can grow.

The female bear eased slowly into the clearing on the south side of the hill. She was seven years old and had attained her maximum growth. Her head and body together were over seven feet long and when she stood square on her pillar-thick legs she was over four feet at the shoulders. She weighed about seven hundred pounds. Since it was still mid-May her lustrous golden bronze coat was prime. Only a few—perhaps 5 or 6 percent—of the bears on the Island had coats of this color. In another six weeks she would be ragged from the shedding and rubbing that would continue through August. Still fresh from her winter's rest, however, her fur was thick and lush.

Her humped shoulders distinctive as she stood, she rotated her head on her short, muscular neck, her nose pointing straight up. Her small, close-set eyes could tell her little, but her keen ears and sensitive nose would report most of what she had to know. The unrelenting demands of survival had sorted these things out over thousands of centuries of evolution.

Her ears, small, round and erect, set far apart on her broad skull, twitched as a downy woodpecker chinked metallically in a tree nearby. Her lips rolled back and she woofed hoarsely. Her somewhat pointed jaws, loosely articulated for grinding vegetation, moved easily and she stooped to graze on some meadow barley, then shuffled forward to where the favored bluejoint grew. These preferred grazing plants and others as well, beach rye, the sedges, nettle and seacoast angelica, had brought the sow down from the mountains for the first phase of her spring feast.

She stopped often to listen, for she was seeking more than food. The preceding summer she had had cubs of a previous mating still with her and their sucking stimulus had inhibited ovulation. Free now, and alone once again, her seasonal estrus had begun. Descending ova were ready for fertilization and the sow's behavior would be dictated by the compelling instinct to mate again. Nature, intent on the propagation of her wonders, arranged such matters carefully.

Abruptly the sow stopped grazing, raised her head, sniffed, woofed softly, then bawled. Her call ended in a whine that mixed intricately with a loud chopping of her jaws. Although she could not pick out his shape she knew a boar stood back among the trees examining those of her secrets that could be wind-borne. The mating play had begun and would not be concluded until nature had assured herself of another generation of brown bear cubs. These two giants had survived many dangers and difficulties; the price of their survival was more of their kind.

The female stood in the middle of the clearing, woofing hoarsely and occasionally whining. The male moved toward her cautiously. He had been following her for hours. His nose told him she was in heat and ready to break her solitude. Still, instinctively, he knew that if he was wrong and if this was a sow with cubs, he could expect an explosive reaction to his approach. Males are often cannibalistic toward cubs, their own as well as those of other boars, and females can be quick and savage in defense of their offspring.

The big chocolate boar, more typical of brown bear

color than the bronze sow, tested the wind continuously as he shuffled toward his prospective mate. If she was the sow who had laid down the tantalizing trail, her standing fast in the clearing was a good sign. If she was a different animal, it was a dangerous situation, because it meant she had decided to fight. Few females will stand up to a boar, but those that do have the advantage of a determination more fierce than a male's hunger for cub flesh.

Satisfied at last that this was the sow he had been following, having sorted her out from the scents of other bears that had passed through the clearing earlier in the day, he quickened his gait.

As they came together they woofed and rubbed noses in a gesture surprisingly gentle for such formidable animals. In a few minutes they were feeding side by side in a most sociable manner. In fact, only on this occasion would either of them seek out or even tolerate the company of another mature bear. Between matings they lived solitary and short-tempered existences.

There was no breeding that day or the next, the time being spent in companionable foraging. Late the following afternoon, however, the male began to exhibit more precise interest in the sow and she was obliged on several occasions to plunk her ample bottom down hard to stop the rude intrusions of his nose. He was becoming more persistent. She was slow to respond but the critical business which they were about had softened their dispositions and there was no brawling. That would come later.

On their second night together the pair did not bed down apart as they had done the night before. There was a growing intimacy and as they lay close in the dark on the side of the hill they nibbled at each other's lips and occasionally slapped each other with ponderous paws. Their dark brown claws, recurved, strong and ever available, had grown long during the winter denning and were not yet worn down. They were not brought into play, though, and the slapping was good-natured with broad, plantigrade feet. Paws that could smash small trees with a single blow were used for caressing, and so the night passed.

There seemed to be an understanding reached during the playful hours of the second night. On the morning of their third day the sow submitted easily as the great boar covered her. The surrounding woods echoed with the wonderful range of their voices. They came apart after a few minutes and began to feed on tender spring plants almost immediately. Later that afternoon, when the boar covered her with his great bulk a second time, they remained locked together for nearly an hour. Erectile nodes blocked the vagina and kept the precious sperm from being lost.

During the days that followed they copulated several times more. When the sow showed signs of wanting to break loose, the male would plant his paws in front of her hips with his head lying along her neck. Her wriggling was to no avail and only when she began to whip her head back and forth and make violent chomping sounds with her jaws, only when he could sense her mounting anger, would he release her. By the end of the first week she was far less tolerant of the male's insistent appetite.

Early in the second week the female watched with marked indifference as the boar, an experienced warrior of eleven breeding seasons, chased a smaller male away after giving him a vicious beating. The pair had been feeding apart for several hours each day since the end of the first week and the young male had approached the bronze sow as she grazed alone on an easy slope. The sudden appearance of the older boar on a ridge above startled the less experienced male, who was soon routed. The hunter who was to take the smaller boar's life the following year would wonder how he came to be missing an ear.

The big male returned to the female's side to find no recognition of his valiant deeds. Whether to reprimand her, or just because the fray had shortened his temper, the boar cuffed his mate rather too violently and she ran off and sat down among some nearby trees to sulk. It was two days before he saw her again. When they met she allowed him to mount her for the last time. During the ensuing ten days they met often, fed together for hours on end, and even bedded down close to each other on several occasions, but their sexual interest in each other had all but evaporated. At any time now their innate need for solitude would repossess them and they would drift apart permanently.

In the middle of the fourth week they met for a brief hour of feeding, but that part of them that demanded solitude had gained dominance over their sexual pattern, and they dissolved their union without ceremony.

The next morning found the sow moving in a westerly direction. Two fertilized eggs inside her uterus had already started to develop but would undergo a dormant stage before becoming implanted. Although the month was June it would not be until December that the embryos, potentially great beasts that could weigh almost three quarters of a ton, would be three quarters of an inch long. Since nature had ordained that all bear cubs around the world be born during the last week in January or the first week in February, the delay in development was essential to the schedule.

Alone, now, and hostile to all other bears, cubs and

adults alike, the sow moved off and began feeding in earnest. The year was 1950 but it could have been any one of the four and a half million that have passed since the Pleistocene, when the species emerged. Its origin is traceable to a time twenty million years ago when *Hemicyon*, part dog, part bear, stalked lesser Miocene fauna. The more direct ancestor of the species was *Uasavus*, a wolf-sized bear of Europe as it was fifteen million years ago. A few million years later, in the Pliocene era, *Ursus arctos*, the European brown bear, was on the scene and from it descended all the brown bears and grizzlies whose ranges circle the globe in the Northern Hemisphere. When man was still only a vague potential in an ape's loins, the basis of the bears' mating ritual was already millions of years old. The precision of its formula stems from that antiquity. Although the bear is an intelligent and adaptable creature, in matters as critical as this neither of these two qualities is required. Nature does not trust such basics to choice. In mating the bear is guided by instinct; its behavior is rigidly controlled.

·   ·   ·

There was no specific plan to the sow's general movement toward the west and south. She had moved down into the valley from her winter denning site, mated, and was now continuing her wandering without conscious concern for her goal. She was biding her time before the start of the salmon run that would take her to certain streams in the area. Spring was passing into summer with its inevitable battle of white versus green and brown. White would lose and retreat to the highest hills in the east and north. The blue lupine, the white windflower, ragwort, four species of orchids, yellow violets, blue irises, flowers shaped like bells and others like stars, flowers sweet and some with poisonous roots, grew in wild profusion. Color was creeping back into the land and overhead the activities of the birds became frenetic.

Without regard for the havoc she created, the sow wandered from larder to larder. Birds challenged her whenever she passed a nest or brushed against a favored tree. Year-round residents, the magpie, black-capped chickadee, the varied thrush and the crossbill, cocked their heads and worried about her size. Summer visitors, violet-green swallows, the hermit thrush, pine grosbeaks, redpolls, and dozens more that had been goaded into their perilous journey to the Island by a fury and drive they could not understand, discussed her every move. As she moved by day, slept by night, she was abused and cursed by a shrill chorus of countless voices. Several times she was mobbed by a mass of swallows who flew at her in a steady stream. Bewildered and frustrated by their dive-bombing tactics, she shuffled off in sullen dignity.

East of Deadman Bay the sow reached the coast. She argued with some raucous gulls and took possession of the carcass of a Pribilof fur seal. An old warrior of many seasons, the bull had sickened at sea and wandered too far to the east. Infested with hookworm and doomed to die, he had crawled ashore and remained half alive while a dozen tides came and went, first washing over him and then leaving him wedged between sea-battered rocks. The gulls had started to feed on him before he was dead. His eyes were taken first. He had lost both the will and the strength to resist, yet the power of life within him was too strong to allow an easy surrender. The natural order of things began drawing his chemistry back into the cauldron while he still lived.

The sow drove the gulls away and began to feed after ending the seal's misery with a single blow of her great paw. That night she bedded down in a clump of trees at the head of the cove and reclaimed the carcass the following morning. A hundred gulls moved among the rocks nearby and hovered overhead, maintaining their lament. Her indifference seemed to anger them further. Beyond the tide pools more gulls floated on the momentarily gentle swells, blue, gray and white corks, animated and shrill. A pair of bald eagles perched like sentinels on a nearby tree. Despite their aloof and perhaps noble appearance they hungered for carrion no less than the other birds.

As summer approached, the sea birds along the coast increased in numbers and variety. Summer visitors, the common snipe, rock sandpiper, mew gull, black-legged kittiwake and Arctic tern, added their endless movement and noise to those of the permanent shore residents, the glaucous-winged and herring gulls. The magnificent golden eagle, a summer visitor only, appeared and matched aerobatics with the bald eagle who was king year-round. Loons, grebes, albatrosses, shear-waters, petrels, cormorants, geese, ducks, whistling swans, sandhill cranes, murrelets, puffins, scoters, oystercatchers, some resident, some purposefully present, some transient, and others accidental, appeared by the thousands and turned the great Island into a vast aviary. Short-tailed weasels and red foxes worked their way along ledges and into brush piles to harry the ground nesters, and destroyed thousands of eggs and fledglings. The Island feasted on its own abundance. A billion times a billion food chains took microscopic form in the soil and the sea. On land the great bear was the largest creature and in the sea it was the whale, but each depended, ultimately, on animals too small to be seen by the naked eye to supply the chemicals upon which the whole complex scheme of life was based.

4

The Monarch Emerges

From
Monarch of Deadman Bay

A cub emerging into the environment in which it will play out its life story finds a "place" already solidly established. The ability of that individual animal to fit in is really what its life is all about. That is, adapting as an individual to a place that will not adapt at all is what determines if a particular young animal will survive, mature, and reproduce. True, there is an undeniable quotient of luck involved in the survival of any organism, but adaptability is really the key. If an animal cannot establish the space it needs in the place where it is to live, nature really doesn't need that animal to be able to reproduce.

For the writer of natural history it is important to get that young animal and the reader together as soon as possible so the reader can understand what the young animal is being called upon to do. Also, it is yet another chance to bring in elements of the setting. Bushes and other growth can be named without sounding like a seed catalog, and weather patterns can be established. The naturalist/writer must grab every opportunity there is to keep the story moving forward while imparting as much data as possible.

Usually Latin and Greek names are not used in the body of the text although they can be. It is all a matter of style and there is no best way of doing it. Any elements of "gee whiz" that can be inserted increase the author's chances of keeping the reader interested. After all, there isn't much point to the exercise if the reader drops off by the side of the road.

The world of change into which the cubs emerged was already far advanced. Cubs of previous seasons, the yearlings, sows that had not bred the summer before, and the unpredictable males were about. The influx of bird life and the offshore flow of marine mammals heading for the newly liberated Arctic Ocean pastures were in progress. Rain was a daily and sometimes hourly occurrence and the ground underfoot was mushy. Spring was unmistakable on all sides and summer was on the way. Her advance scouts were everywhere.

Shortly after leaving the cave, ahead of her cubs and extremely alert to the possible appearance of a mature bear, the sow began to eat cathartic grasses and herbs and quickly voided the black, resinous plug that had blocked her intestinal passage. Her feet were tender from the long period of inactivity and she limped slightly. During the first days she stayed close to her den, eating what she could find on the higher slope. Not at all unlike a cow, she would take a mouthful of grass and crop it by a slightly abrupt lift of her head. She was still fat but would lose weight rapidly during the first two weeks. Following that she would again begin to lay on fat against the needs of her coming sleep.

This concentration on food is typical of bears. The demands increase as spring progresses into summer and the sow, never a fastidious eater, took whatever she could find. While she might consume surprisingly little for so large an animal at any one feeding, her meals were so frequent as to be almost continuous. The total volume of food consumed was larger than might be suspected by the casual observer.

Seeking the tender pooshka, or wild parsnip, the sow would grasp a mouthful of vegetation and plant her front paws firmly on the ground. With a convulsive movement she would thrust backward with her body until a clump of sod tore loose. Turning it over with her paw she freed the roots, up to a half inch or more in diameter, and slowly ate them. In her quest for these tender morsels, and for grubs and beetles as well, she turned over whole areas of the hillside until it looked as if it had been plowed by a drunken farmer. Food-getting for a bear is more a matter of drudgery than of reliance on keen senses. Having given the bear a varied appetite, having delivered it from the agony other predators know when game is short, nature has either taken back or denied altogether the razor-edge alertness that wolves, weasels, and cats must have to survive.

At regular intervals the sow returned to her cubs, for their feeding demands were no less insistent than hers. Unlike their mother, however, they could accomplish nothing on their own.

Often, as she worked the fields close to the mouth of the cave, she would leave her cubs at its entrance, but she was never out of range and she constantly tested the wind for signs of danger. When she came to them they whined eagerly and climbed over each other to get at her. She would sometimes lie on her side and watch them feed, making the softest of satisfied sounds. At other times she would lie on her back and move her hind legs rhythmically as they tugged and gorged. And at yet other times she would sit square on her bottom with her back against a tree or mound and place a paw on the back of each cub. With her hind legs thrust out in front like a comical old woman she would point her nose straight up and slowly rotate her head as if to exercise a stiff neck. The cubs thrust hard with their hind feet and shuddered with satisfaction at what she gave them. Always she was tender, always alert. Her life was divided between feeding herself, and through herself her cubs, and worrying about their safety. There seemed to be no other forces, no other concerns.

The cubs grew daily. Their emergence weight of fifteen pounds would have to increase to a hundred pounds or more by mid-autumn. By the late fall of their second year they would weigh as much as four hundred pounds. A difference in weight between them would not occur until about their fourth year. For the moment, there was little to distinguish between the two. They were liver-gray in color but it was impossible to predict the tones they would finally achieve. The genes they inherited from their parents had been too confused over the preceding generations by the influx of brown bear color variation to take a predictable form. Since no survival factor had existed in any one tone before man arrived there was no particular trend. Before nature can make that miraculous adjustment man will almost certainly see to it that the bear is extinct.

As the cubs' size and strength grew and as their coordination improved the sow increased the length and duration of her excursions. Calling to them and constantly bolstering their confidence with the sounds she made, she took them further and further away from the cave. At last she began keeping them away for days and nights at a time, always bedding down before dark in the deepest cover she could find. Their demands on her never faltered and their treks were often interrupted for a feeding session. The further they moved away from their den site the more alert she became. She seldom relaxed for more than a few minutes at a time.

One afternoon as the family was edging down through a clearing between two rings of stunted alders that girded a hill, the sow stopped short and rose to her hind legs. The movement was smooth and effortless. Straining against the inadequacy of her vision she moved her head from side to side. The cubs came tumbling up against her legs and began to frolic. She issued three rapid, harsh commands and in a comic imitation of their mother they attempted to rise up to see what had caught her attention. The longer she held the position the more nervous the cubs became. They sank to all fours and moved in close against her legs. The female cub began to whine and again the sow grunted peremptorily. She was listening to the winds and sampling their chemistry. She sensed another bear in the vicinity—and it was close by.

On the lower portion of the slope, another sow stood among the alders and stared myopically up to where the bronze female towered. Victim of a natural freak, this bear had *four* cubs huddled by her legs. This extremely rare occurrence does happen from time to time and the sows involved are generally all but overwhelmed by the ordeal. With so much more to do, with so much more to worry about, their whole attitude is one of profound bewilderment.

A small current of moving air that had begun at sea and picked its way across seaweed-covered rocks, through patches of brush and trees, was working up the slope. The energy behind it was reinforced by other currents from over the surface of the water and it flowed and rippled across the clearing. It passed the sow in the alders, snatched away her secret and eddied past the female on the slope peering down, alert but uninformed. Instantly, the bronze sow located the intruder in the valley. Her sudden head movement and grunt caused the stranger to move, and to shift her position ever so slightly. The bronze sow was able to detect the movement and determine her shadowy outline. She gave a sharp bark and lumbered two steps forward on her hind legs before dropping to all fours, facing downhill. Her cubs were already on their way up to the ridge. They bawled in terror as they ran.

With front legs stiff, each step jarring her great frame, the sow hurried down the slope.

In the alder growth the other female, too, had gone to all fours and, determining that her cubs were well concealed, started out into the open.

The two sows faced each other over a distance of a couple of dozen yards and circled slowly until they were on the same level. In a kind of displacement activity, as if to relieve the unbearable tension that had been mounting, the intruder stopped and pulled free a mouthful of grass. Jerking her head up she quartered away and stood with her head turned to the side, looking in the direction of her opponent with the grass drooping comically from the corner of her mouth. In an imitative movement the bronze sow did the same.

Then, without warning, after having given it all the thought of which she was capable, the bronze sow charged. She hurtled across the intervening yards and caught the intruder in the shoulder as she turned and half rose to bring her great forepaws into play. They slapped ineffectually as she was rolled over twice by the weight of the impact. Her reflexes had been a beat too slow and the blood flowed from an open wound where the sow had sunk her teeth.

The momentum of her charge carried the bronze sow well beyond her target and when she pulled up and whirled about to charge again she was struck by the intruder barreling down on top of her. She felt a terrible, stunning shock as a paw as large as a platter with powerful claws spread wide and angry descended with the full force of half a ton behind it. One of the bronze sow's cheeks was opened and her teeth showed through the wound. Again she charged, snapping furiously, but the intruder had already begun to retreat. She caught up with the darker female and managed to sink her teeth into her rump before she vanished into the brush. The crashing of her great body sounded as if a truck were hurtling through the growth.

The sow patrolled the edge of trees, coughing and grunting. She didn't dare enter the thicket with an opponent so aroused and with the benefit of cover. The air currents between the trees could not be trusted and her eyesight would be all but useless.

The bronze sow's two cubs and the intruder's four had witnessed the battle huddled in two groups a hundred yards apart. They would have played together had they been allowed, for they were still endowed with a social sense that enabled them to tolerate their littermates. They would lose it in time, though, and were learning the lesson of distrust that would stay with them as long as they lived.

Both females bedded down almost immediately after returning to their cubs. They were no more than a hundred and fifty yards apart in the two groups of alders that bounded the small clearing. Throughout the night they both remained awake, sniffing, listening for the sound of any movement. On several occasions each moved to the edge of the trees and stood facing each other, although neither could know for sure the other was there.

On the following morning the sows again spotted each other. They did not clash, although some short charges were made by each as gestures of threat. They drifted apart after a few minutes and did not see each other again for several hours, when once again they came within sensing distance of each other. Several defiant movements were made, but again there was no direct conflict.

On the morning of the third day, shortly after feeding her cubs their first meal of the morning, the bronze sow moved down to the edge of the trees. There, not more than a dozen feet away, the intruder grazed with her four cubs strung out behind her. The wind was blowing again from the sea and the scent and sound of the intruder carried clearly and unmistakably. The sow sank back on her haunches and sorted out the messages. With a wild roar, almost a scream, she burst from her cover. The four cubs scattered but one was too slow. Snatching it up in her great jaws she ended its life with a single snapping action, dropped its small body and spun again to re-enter the woods where her own cubs were wailing.

Whether or not it was immediately clear to the intruder that she had lost her smallest cub we cannot know. Her remaining three were running and tumbling down the slope in abject terror. The charge of the great bronze sow out of the brush so close at hand came with stunning impact. Only their training enabled them to break away from the paralyzing effect of the attack and get away at all.

The intruder spun around, perhaps seeing the body of her cub lying limp and oozing blood, and crashed into the brush after her opponent. Roaring, wailing, grunting, and chopping her jaws, she smashed down brush and with a gesture of wild defiance clubbed a sapling an inch and a half thick to the ground with one sweep of her forepaw. Rising to her full height, her jaws still chopping in anger, the great sow circled slowly, worrying everything in her way. In her passage she destroyed the nests of three ground-nesting birds. The yellow yolks from a dozen shattered shells seeped out and the parent birds circled overhead, bemoaning their loss. Diminutive mammals of several species fled before the onslaught and a mouse nest toppled, spilling its pink inhabitants to the ground. When the sow had passed a weasel emerged and took the little bodies before the female mouse could find them.

The furious charge of the intruder into the brush was to no avail. While she beat her way through the bushes and between the trees the bronze sow and her cubs had vanished over the ridge above and were close to a mile away when the intruder emerged grunting and coughing on the downslope side to sit wailing beside her dead cub. She left the valley that day and never returned.

As if her cruelly violent deed had reminded her of the danger that surrounded her own two cubs, the sow was unusually alert in the days that followed. She was even short-tempered with her charges and their obedience had to be ever more unquestioning to satisfy her. She cuffed them often and bit one on the flank hard enough to make it whimper for several minutes. Thoroughly cowed, it returned to her to be fed and found her forgiving.

The intruder that had come to the valley to lose her cub

remained confused and miserable for days. She never quite realized that he was gone and grunted angrily several times when her commands brought only three cubs to her side. She would look for him and stand bawling when he did not appear.

5

The Wolf and the Squirrel

From
The Custer Wolf

A young animal's place exists not only in geography and topography but in time. For a young animal (and this is true of human beings as well) there are stages, places along the way that I might call milestones if I allow myself a little anthropomorphism. Young animals that pass these significant points can continue on their rights of passage.

For a young wolf a new kill, a trickier kill is such a milestone. For all young predators everything that moves is an educational toy. What a newly emerged cub does with a ground squirrel is play perhaps in human terms, but in wolf terms it is practice for survival.

In a sequence like this one, I get a rotting log, a species of mushroom, a species of butterfly, and a rodent species into the scene. They are all part of the place where the wolf cub lives and to which that cub must adapt. Here I hope I have built the setting slowly without making you feel like you are doing homework or being given lists to memorize.

Near where the wolves lay there was a heap of rotting wood. Melting snow had left little pools in cracks and dents that helped the process of decay. On this rich garden there grew spicy tufts of the velvet-stemmed collybia, the tangy early mushroom of springtime and dead wood. The sticky, reddish-yellow caps with their tawny margins and the firm, velvety stems arose from death but heralded new life and beauty. Attracted by the aroma of the collybia but surely immune to its true beauty were two spicebush swallowtails. Relatively rare so far west, these four-inch butterflies skidded low to the ground, changing from blue to green in iridescent waves of living light as the sun's intensity was filtered and changed by the foliage overhead. Softly coming to rest near the mushrooms, they had become the focal point of another's attention.

The thirteen-lined ground squirrel was nearly a foot long, although four to five inches of that length consisted of unimpressive tail. As the name implies, his body was marked with thirteen lines, yellowish-white alternating with tones of chestnut, the darker punctuated with white spots. His face and underparts were a tinted buff that appeared first one tone and then another as his nervous movements constantly altered his relationship to light and shadow on the floor of the glade. His tail was yellowish-brown, fringed with coarse black hairs, each endowed with a yellow tip. He was as beautiful as he was nervous: a shadow, a dart, a flicker, a spot of life as he stalked the swallowtails. His great oval eyes saw all within immediate range—but they had missed the larger scale.

And so here was a moment in time, an infinitesimal microcosm in a world suspended somewhere between dinosaurs and space travel, a drama not only as old as time but reflecting the very essence of time itself. Here had germinated a seed and grown a tree tall and proud and here it had died after the nature of things. The elements had construed to pulp the wood, flake the bark, dissolve the chemicals and the stuff of its once impressive height and girth. On this decay a fungus grew, the spicy, scented collybia. The collybia, in turn, had drawn to it two mites of life, two wood nymphs in the persons of the swallowtails. In turn, stalking them, came the ground squirrel. His diet, half animal, half plant, had room in its range for these wisps of life, and he, this gentle creature of the daylight hours, became the hunter. Perhaps in justice, the turncoat squirrel was in turn hunted by the wolf cub—not so much in need of food as of experience.

Although he was in his spring weight—no more than four or five ounces—the squirrel had one very important element in his favor: a collection of twenty-two teeth, the front ones honed fine and sharp, ready for cutting. He also had a tenacity as rich and wonderful as that of the wolf cub—he wanted to live. He could not comprehend life, not even self; but in a finer, more precise wisdom, he wanted to maintain it. Here, surely, is another miracle we must acknowledge. Witless ones, unable to so much as begin to comprehend the existence of life, will nonetheless perform stupendous feats of bravery and heroism to defend it.

And thus the wolf cub came closer, not only to the squirrel but to the knowledge that life was not to be easy. Here was food—prey—that had to be taken; and although being a wolf might be the grandest miracle of all, it was not necessarily enough. You had to be a skilled wolf to survive, not just a wolf.

The consciousness of a ground squirrel, in matters of survival, does not deal in subtleties. He is either apprehensive or terrified, there is no other course. As the little animal moved in on the swallowtails, as intent as he was on his game, he was also alert to danger to himself. A sudden shadow could mean a hawk overhead, a rattlesnake striking from a tuft of grass, a fox or weasel, a coyote—even a wolf. For the ground squirrel, danger was everywhere: in all movement, in all sound, and in silence as well.

It was not surprising then that the wolf cub's presence was quickly sensed. With a sharp little squeak and a sudden spring-like action, the squirrel leaped away, spun around and, on his hind legs, faced his foe. He was a burrower, not a climber, and without his burrow he would have to find another hiding place or stand his ground. He looked about. There were no hollow logs, no piles of stones, no mountains of forest debris. The remainder of the dead tree on which the mushrooms grew was too far gone; there was too little left of its substance to provide shelter. No, it was there—in that glade where the sun fell in shafts and painted stripes on the ground—that the mature, life-wise, thirteen-lined ground squirrel must face the immature, clumsy cub. The outcome was by no means certain.

When the rodent cried out and inconveniently leaped away from the spot where he could easily have been taken, the wolf cub, in surprise and confusion, stood up to his full height. Despite himself, as much as he wanted to be a wolf, the best he could do at first was cock his head to one side and whine. He had not faced such determination before, nor had he ever seen so small an animal rear up, face him, and scold so loudly. It was all rather confounding.

The scolding of the ground squirrel, the shrill outpouring of pure rodent invective alerted every animal around.

Songbirds resting the while in nearby brush rose full to the sky and were gone. An owl high in a sheltered place turned his head an incredible three hundred degrees and blinked down. If he had cared—and he surely didn't—he would have cheered the wolf on.

Here in the center of a private and very temporary universe, the white wolf cub became the focus of attention. For the first time in his young life he was on trial; his measure would now be taken. Quite suddenly, although the terms of its understanding were somewhat cloudy, the taking of this prey, the killing of this strange little chattering creature became the most important thing in the world.

The animals he had killed before had been easy. His needle-sharp teeth sank quickly to a vital spot in those small bodies and the struggles ended as suddenly as they began. What could be so very different here? The body was small—although apparently made of a kind of spring steel—and a bite was all that was needed. A simple, quick bite with canine tooth penetrating skull, spine or chest cavity. Nothing more than that, certainly.

One characteristic that marks the rodent is the nature of its strange front teeth. Those four curving, gnawing tools grow for as long as the animal is alive. They never stop seeking a complete circle, for they grow in great arcs that will eventually close and cause the animal to puncture and kill itself—stab itself slowly as it were—if hard and resistant substances are not always at hand on which the teeth can be worn down. A gnawing animal, then, must constantly gnaw in order to stay alive. Because of this dominant fact of life, its four front teeth are always honed to a fine cutting edge, always dangerous to young and inexperienced predators.

And so, without regard for tactics that he would one day

know and use as a matter of course, the wolf cub moved in on the ground squirrel. Stiff-legged, a little hesitant, somewhat intimidated by the constant flow of invective that was being heaped upon him in such shrill tones, Lobo began his approach. His first fateful mistake was in letting curiosity get the better of his kill-instinct: as Lobo pushed his snout forward to smell his foe, the ground squirrel, not at all interested in the fine distinction between this indignity and a fatal thrust, fastened himself to Lobo's upper lip and felt his teeth meet in the middle. With a yapping shriek, the wolf leaped backward, outraged, shaking his head and striking at his face with his paws. The ground squirrel lost his grip and hit the ground, rolled, righted himself, and again faced his foe on his hind legs, scolding furiously. Before Lobo's convulsive reaction to the initial pain had enabled him to shake the little rodent loose, he had been bitten three more times by jaws that moved with the speed of a sewing machine.

Lobo stood off now and studied the dancing, chattering sprite. He shook his head from time to time, splattering blood in all directions as it flowed freely from the wounds on his lip and muzzle. The white fur there was stained red, as were his front paws, and there were splatterings on his chest and shoulders. He was, indeed, sorely injured, but not as badly as he may have looked to the casual observer. An animal defending itself against a great predator, even the young of great predators, works under one great disadvantage. If, in its self-defense, it injures its foe, it doubles the intensity of the assault it must face. A wolf is never so much a wolf as when injured in pride or body. A killer is never so much a killer as when a game is involved.

Before the ground squirrel's teeth met in Lobo's face,

before this gross indignity was heaped upon him, Lobo wanted to kill the squirrel and lay its drooping body before his parents, and then fight for it with his littermates. But now, all splattered with blood, the young wolf knew it was no longer a matter of *wanting*—it was something he *had* to do.

Wary now, alert to the danger inherent here in the foe, Lobo began feinting in the classic manner of the wolf. First he would make a brief, short rush and then suddenly pull up short, leaving the adversary confounded and increasingly less certain of itself. That was his tactic. In, out, in, out, Lobo kept up the pressure. Soon the little squirrel, with a heartbeat and respiration rate very much greater than the wolf's, was fairly lathered into blind fury. Each thrust by the white wolf cub put him off balance, made him adjust and readjust quickly as the wolf withdrew. Never close enough to bite, yet always thrusting, the wolf became a maddening object, not just a frighteningly large one.

Finally, as Lobo made his thrust, the sense and caution of an animal who had already survived one hostile year gave out and the squirrel charged the wolf. Alert now to the teeth of rodents, Lobo sidestepped quickly, turned and brought his great paw down on the squirrel's back. The little animal twisted within his own skin and cruelly lacerated the cub's right front paw, causing more blood to flow.

The two foes quickly resumed their opposing positions, the ground squirrel on his hind legs, dancing to and fro, the wolf cub circling around him, haunch high, forelegs to the ground. Now, however, the cub was forced to stop from time to time to lick his damaged right front paw. During one such hesitant moment, the rodent grasped an opportunity—returning to his four feet, he streaked past the wolf up the rise toward some thicker wood. Although more an animal of open places, without a burrow at hand he sought cover in any form. His victory, after all, could only come in escape; he could not kill the wolf, and thus he sought only the stalemate he had held with life from the day of his birth.

When the rodent turned his back on the cub, however, he had already made his fatal mistake. Even with his injured paw, the cub was fast enough. In two bounds, still awkward and seemingly more frisky than deadly, he overtook the squirrel. With the advantage of coming up from behind his foe, away from those slashing incisors, he quickly closed his teeth, felt the satisfaction of penetrating living flesh, and held a dead squirrel in his teeth by the time they met in the middle. A single well-placed snap had crushed out its life and made the wolf cub the victor. A career was launched there that day.

With blood still flowing from his lip and from his lacerated pads each time he lifted his paw to take another painful step, Lobo limped back up the rise. He reached the small sheltered area where his parents and littermates lay up in the shade. There were no cheering crowds, no voices to call out "well done." There was, though, a look of approval, a kind of silent acceptance. Once again a cub had done what was right, what was expected and necessary if he was to survive. Wolves are not human beings who must run around reassuring each other, thumping each other on the back; these great wild predators do exactly what they have to do, or they die. The fact that they survive at all is their reward.

The white wolf cub stood at the edge of his family group and looked around. The great tan male sat motionless on the higher ground looking down with his yellow eyes. His silver mate lay in the shade, languid, serene, her softer eyes resting on the white cub as well. Three small bundles of coiled springs and intermeshed gears huddled nearby awaiting a signal giving them permission to explode. Screwed down to the ground by an invisible force, they needed only to be set free to spiral upward and outward in all directions and resume the games. Lobo dropped the limp ground squirrel, let it slip from his jaws. He sank back on his haunches and whined softly. Then, there being no apparent objection from either parent, the cubs let loose. At once they were upon the squirrel, over and around Lobo, and were gone into the brush, tearing after each other, lusting for the prize that Lobo had brought home.

Still under the eyes of his parents, Lobo sank full to the ground and began working on his front paw. From time to time he looked up. The eyes of his parents were never off him, not even after he had fallen asleep to relive the chase and the kill.

· 6

A Strange River Home

From Panther!

There are places on this planet that are truly unique. In one sense every place is unique because it has its own texture, but technically a place like the Everglades has no counterparts. It is a vast river fifty miles wide and on the average less than six feet deep. When a nature writer spends a lot of time in a place like that (and I did!), there is a temptation to devote whole chapters to the wonder of what has been seen. But readers who think they have picked up a story may not be willing to read chapters of the author's enthusiasm for a place. Readers want to get on with what happens there and, before too many pages have passed, with how the hero or heroine of the book fits in. It is hard to resist the temptation of a setting as rich as the magnificent Everglades. Writers are by nature usually (to some degree at least) undisciplined. They are in it for a lot more than the money. One of the greatest rewards of being part of show-and-tell is the luxury of being able to indulge one's enthusiasm for places like the Everglades.

The land where the female panther roamed is a strange one, a unique subtropical wilderness of nearly twenty-two hundred square miles. Time and again, more times than we shall probably ever know, the land there has been reclaimed by the sea. Each time the water rolled back again, another layer of sediment was left behind to cover the rocks of the peninsula. Slowly, the porous land evolved as an enormous basin sloping toward the south and west. For centuries, rain and drainage from Lake Okeechobee to the north were kept from seeping down through the underlying limestone by a cover of marl and peat. The great spillage from Okeechobee's depressed southern boundary did not form into rivers as it ran to the sea, but flowed across the land. The limestone outcroppings that appear to the north and east are now the foundations of southern Florida's cities, and where they appear in the Everglades itself they are the hammocks upon which the hardwoods, ferns and air plants grow.

The elevation within the Glades varies from sea level to ten feet. Even these very slight variations are enough to create the several distinctly different habitats into which the region is divided: willow swamps, hammocks, sloughs, cypress jungles, saw-grass flats and pine flatwoods. The entire ecosystem is tied to water supply, for the Everglades is first and foremost a land of water. Without a steady supply the animals there sicken and die and the sky is blackened by the smoke of grass fires. Here the land must be wet or it dies.

The Everglades is a transitional zone, linking the continental floristic provinces of North America with that of the West Indies. The exchange of vectors, water, wind and birds, with the islands to the south has resulted in a mixed system unique on our continent. The saw grass, often ten feet high, is closely related to sedges found in New Jersey bogs, while the mastic and gumbo-limbo trees are essentially West Indian species. Botanically, the Everglades is a melting pot.

Plant life competes with itself no less than do animal forms in the Everglades. Competition has driven many of the orchids and other air plants up into the trees and forced the wild mangroves out to the edge of the sea, where they alone can survive ankle-deep in the warm brine of the Gulf. The coastal features of the area are products of the mangrove's retreat in no small part. The seeds of the red mangrove, the island builders, germinate while still aloft and often hang on until the embryonic root is a foot or more long. If the ground is exposed when the seedling falls, it plants itself. If the tide is in or the water high, it floats root down until it reaches exposed land, where it establishes itself in the ooze. It grows rapidly, with arching roots forming a thick tangle. Floating debris and animal colonies are caught and established until the beginnings of a new island are formed. In effect the trees have retreated from the fierce competition of the land and gone to sea, where they can form their own world.

It is a rich land and life is produced in incredible profusion and variety. Fifty species of mosquitoes inhabit the area. They have evolved so that the competition between them is limited. They have specialized their habits so that they are active at different times of the day and night. A third of all the ferns found in the United States are found in Florida; sixty of these hundred are of tropical origin and are limited to the region of the Everglades. The area is their single toehold in the United States. Three hundred and twenty-three species of birds have been recorded in the region, two hundred and twenty-seven occur regularly, eighty nest there. The native butterflies and skippers number one hundred and forty-two, and the amphibians sixteen. The reptiles number fifty, including such dramatic forms as the alligator, the few remaining North American crocodiles, and the eastern diamondback rattlesnake, potentially one of the dozen or so most dangerous snakes in the world.

Ten species of whale and dolphin have been found in waters rimming the peninsula and thirty-seven mammals inhabit the land, the most dramatic of these undeniably being the cat they call panther. Elsewhere the same cat is called mountain lion, puma, cougar, screamer and catamount, but in Florida the name is panther, or sometimes "painter" in its corrupted form. The female specimen that mated that troubled night, one of perhaps two hundred-odd surviving panthers in the region, weighed one hundred and five pounds. From the tip of her nose to the darkened end of her three-foot tail she measured a total of exactly eight feet. She stood an even three feet at the withers, two inches higher in the rump. As is generally the case with Florida panthers, she was darker, richer, a more distinctly reddish-brown than cats of her species found to the north and west. The mountain lions of Arizona, Colorado, California and Montana are pale in comparison to the short, stiff rufous hairs that constituted most of her coat. Her ears, like the tip of her tail, were dark and shone from deep cinnamon to black, depending on the light. Her belly was white, as was her rump, to the first hind-leg joint. Her upper lips near the sides of her nose, her chin and throat, were nearly pure white, with pinkish-buff overtones on the underside of her neck. The sides of her neck were a cinnamon-pink buff, as were her feet. The light played tricks with her colors and what might appear at first glimpse to be a fairly uniformly colored cat was in fact an intricate blending of subtle, muted tones.

In most ways, though, she resembled all of the other mountain lions of North and South America. She had large paws armed with devastatingly sharp claws, five on each front foot, four on each rear. They were an inch long and gave her enormous gripping power. She was graceful, sleek and strangely mysterious, a perfect specimen of her kind in appearance and behavior. She was an outstanding example of an outstandingly successful species. Her kind has the greatest north-south range of any species of cat in the world. They are found today, as they have been for untold thousands of years, from almost the very southern tip of South America up into Canada. They are found in deserts, swamps, mountain valleys and deep forests. Like all of her species, the panther who bred that night less than forty miles west of Miami Beach was sly, secretive, adaptable, inventive and sure. She was a consummate stalker and still hunter, a climber capable of leaping fifteen feet or more straight up onto the trunk of a tree, and a sprinter. On flat land she could bound forward in twenty-foot leaps; from a stand-still she could broadjump nearly fifteen feet. When necessary she could swim with speed and determination. It was a combination of her appearance and these qualities that made of her a coveted big-game trophy throughout her species's range. These same qualities had the power to engender fear and resentment. It was well known that she could, despite her small size, kill a fifty-pound colt with one pounce and carry it for three miles before tiring. A thousand-pound horse was hardly a challenge for her and she could, if necessary, drag it over a fence or into a ravine. Although her speed could be accounted for by her form, her strength seemed hardly possible for so small a cat. She was, after all, only a quarter the size of a tigress. Her secret lay in her internal structure. She was a completely economical combination of bone, sinew and muscle. Nothing was wasted, nothing was extra and no important elements were missing. In thousands of generations nature had devised a perfect killing machine, a swift, sure, fiercely determined feline predator whose intent on staying alive was matched by her ability to do so.

The Everglades was just one of the hundreds of specialized habitats in which the panther could have been at home. But thousands of years earlier her ancestors had discovered this place, found it rewardingly stocked with a variety of animal foods, and here they had prospered. Her kind was now a balancing element in the ecology of the area; she had a job to do, for her killing was a part of the natural scheme. Except in rare instances the animals on which she preyed had a birth rate to accommodate her depredations. The Everglades was lacking nothing of what she needed. Sleek, sure, generally silent and excessively shy, she was the great trimmer, the unrelenting reaper, the great cat of the Florida southern swamp.

7

People of the Panther's World

From
Panther!

For the very pure among us, the species we belong to doesn't figure in a "nature" story, but that is, of course, nonsense. Very much a part of the life story of most of the species of interest to us is how well people can adapt to them and them to us.

In a wild place it is quite proper to introduce human beings, and if the place is wild enough, wild people are more than proper. They are useful for establishing mood. Of course, people had visited and utilized Bitterroot Island, but what kind of people? For the panther to be seen against as appropriate a backdrop as possible, the human elements could be as wild as the author's imagination. Writers can bring together many stories they have heard and find a place for them. By including human beings as elements in wildlife places, some good lore can be recorded. It is only stretching a point a little, generally, to move people around on a small map and in a relatively short time period. That was all that had to be done here and lo, the female panther had her island with its human history blended into its natural history.

Men had known Bitterroot in the past. Nearly two thousand years ago a band of Tequesta Indians settled below the ridge in the southern end and collected birds' eggs there and hunted deer. They were invaded by, but finally repulsed, a marauding band of Calusas, but not until seven of their own best men were dead. They abandoned the island after that and it remained apart from human events for nearly three hundred years. It was then that Cha-a-na-ra-tee, the ancient and bilious leader of a splinter group of Timucan Indians in northern Florida, sent an expedition into the south to see what was there for them. The warrior that was leading the expedition, Ta-ra-ta-na, succumbed to a rattlesnake bite received on Bitterroot the very day they landed there and the expedition headed back toward the north. Only three of the original eleven men ever made it back at all and Cha-a-na-ra-tee, a less than gentle soul at any time, had them put to death. It was never quite clear to their families why they had to die, but that wasn't a time in history when leaders had to explain themselves, particularly to the kin of executed men.

By the time fourteen hundred more years had passed, the fate of the Glades Indians, to which the Calusas and Tequestas belonged, was sealed. Raiding bands of Creek Indians retreating before the onslaught of the whites in the north were moving south. Later to become known to white settlers as Seminoles, these Creeks were too highly organized for the loosely confederated Calusas, who had themselves long since ridden roughshod over, and submerged, the Tequestas. The Calusas were driven further and further inland away from the desirable lands along the coast and a band settled on Bitterroot. The Calusas had known the Everglades, of course, had explored its reaches by canoe. But few settlements had been attempted and those were temporary. Now a small band of thirty-seven Calusa men, women and children tried to make Bitterroot their home, but a local Seminole chieftain, his name now lost to us, sent his warriors against the village. The Calusas had their bows and arrows and their *atlatls*, or spear-throwers, but they no longer had their spirit. They died to a man and again Bitterroot was abandoned. From time to time after that Seminoles came there to collect turtles, which abounded, and to hunt near the twelve pools along the southwest shore. It was a rich land and one well known to the newly dominant tribe from the north. But in no sense was it an inhabited island.

Shortly after the beginning of the nineteenth century a sixteen-year-old boy by the name of Sammy Goodbody killed his best friend in an argument, the subject of which has never been recorded, and escaped the law by hiding out on Bitterroot Island. When he was about eighteen years old the illiterate youth canoed over to a settlement on the Shark River and in the dead of night "took hisself a cracker girl" of fourteen years, carried her back to his island and told her she was then his wife.

Sammy's whereabouts remained a secret for nearly thirty years, until 1831, when two deer hunters who had announced their intention of heading out Bitterroot way to try their luck failed to return. Two brothers of one of the men who went out to search for them also vanished. Twelve men finally mounted a consolidated search party and found the four missing men hanging from a tree by their ankles. They had been beheaded. The men located Sammy, by then a gross giant of over three hundred pounds, and hanged him from the same tree where he had tortured and killed the innocent men. The first three ropes they used broke. The fourth held. The girl that Sammy had made his wife was quite mad when they found her huddled nude in a dark corner of Sammy's unspeakably filthy hut and they had to use considerable force to get her into one of their boats. One of the men was badly bitten by her in the struggle and later had to have his hand amputated and very nearly lost the rest of his arm as well, from blood poisoning. It was rumored that her teeth were so sharp and her bite so bad because she and Sammy had eaten the children she regularly bore him. The girl, who was then not yet forty-five, but who looked like seventy, died six months after being taken off the island from an unspecified "ailurment." Sammy and his child bride, whose name to this day has never been discovered, were the last full-time settlers on Bitterroot Island. From the very early 1900's on, small cabins were built for weekend use by duck and deer hunters, but they can't be considered settlements as such. Bitterroot, when Panther arrived, was a well-known but uninhabited island.

8

Trapped by an Ancient Foe

From
Panther!

People are not just a background against which animals play out the dramas of their lives. Very often people are antagonists (far less often protagonists). They can become devastating elements in an animal's life, its place.

The old derelict cabin in the Everglades in the episode that follows is a place unto itself, a part of a larger place, the 'Glades, and Doc Painter is a part of each, part of the drama, part of the danger. Men, of course, built the cabin. Doc Painter alters the cabin in preparation for what follows, but Doc's relationship to the panther's world is larger than that, far more significant. Doc is a predator, and predators like the cat are no more free of predation than they are of parasites and weather caprices. Doc, a part of the panther's place, now preys on the panther.

Panther stayed pretty much to the central ridge after his return to Bitterroot. He was never more than a few-score yards from one or another of the twelve pools. He came down off the ridge just at dusk to take some small birds or an occasional raccoon, wading through six-to-eight-inch-deep water to do so. As a result, it was over a week before Doc Painter found unmistakable signs that "his" panther had returned. By that uncanny sixth sense experienced trackers use to confound the uninitiated, the eighty-year-old man recognized Panther's footprint the way a less skilled man might have been able to recognize his face, or his whole form.

The old man was too shrewd to make his move until just the right moment. He knew how to play the cat-and-mouse game as well as Panther. In addition, he understood it. It would be at least several weeks, he knew, before Panther would wander again. He had all the time he needed without rushing things. His timing and his anticipation of Panther's movements had been flawless so far. In the contest, as far as it had gone, human intelligence had been the determining factor. However deficient the intelligence of the ancient illiterate might have been, it far, far exceeded that of any cat that had ever lived.

As the old man could have predicted, Panther did not return to Bitterroot the same cat he had been when he began his wandering. His first odyssey had marked him Each new animal he had killed, each new piece of land he had hunted, had taught him something. And, too, he was now a mated cat. The irony of his having fathered no cubs did not reflect on him at all. He had gone through the motions and he was now a different kind of cat.

Doc Painter was able to detect and perhaps even sense the changes Panther had undergone as soon as he began tracking him. He would find the trail he was following suddenly vanishing from under his nose and he would nod with appreciation. He would smile and look up into the trees overhead. He was seeing "his" cat develop, and he was strangely proud of him. Had he been a younger man, he would have scaled the trees around the end of the trail to find the marks where Panther's claws had fought for purchase. If the old man had climbed a tree or two he would have been likely to find the cat itself crouching low against a heavy branch nearby. Panther was watching Doc Painter and was more than a little interested in the old man's unaccountable behavior. While the human hunter was adept at reading signs, the cat he sought had the sense of smell, the sense of hearing and the ability to conceal himself that gave him the distinct advantage. The cat read the man as often as the man read the cat.

Panther made very few kills that Doc Painter did not soon come to know about. He developed no routines that the old man didn't disrupt by discovering and then, eventually, anticipating. It was becoming a war of nerves and the wise old panther hunter began to suspect that it had gone as far as it should. Any further, he reasoned, *"an' that old panther's gonna up and quit Bitterroot. No sense in pushin' 'im too far."* In a very real sense the relationship a man establishes with a wild animal is like the link he forms with another human being. It is fine, it is critical and it can be easily destroyed. Somehow even the old man understood this, not instinctively but consciously. It had accounted for all the success he had ever known in his life.

For two days the old man stayed away from the hammock, but when he returned on the third he carried a small medicine bottle with a cork stopper. Later that same afternoon Panther came across the first drops of the lure the hunter intended to use as the instrument of his downfall.

Only a few drops had been spilled, but Panther caught the scent and moved down to investigate. He located the rag and rolled on it, sniffed it deeply and noisily, and rolled on it several more times. He picked it up in his teeth and carried it off with him, dropping it from time to time to sniff it and roll on it again. Panthers are no less susceptible to oil of catnip than house cats. Of course, their size makes them appear even more absurd.

The scent soon wore off the rag and Panther abandoned it. It was five days before he found a second lure and treated it in much the same way he had the first. This time, though, the lure was a torn doll and it reeked of man-smell. Still, the catnip was strong and after only slight hesitation Panther ignored the man-smell and abandoned himself to his heady drunken games. The doll was all but completely destroyed before Panther finally left the remnants beside the trail and set off to hunt.

Four times Panther was lured; four times he responded without harm befalling him. He had gotten used to the association of the man-smell with that of the catnip lures. He no longer found it particularly disturbing. On the fifth occasion, when he caught the scent as he passed by an abandoned hunter's cabin, he swung toward it without breaking his pace. Coming upon the scent unexpectedly had become something of a habit, which was exactly what the old man intended. Panther's normal guards were down.

The rich and unmistakable man-smell that reeked throughout the cabin's interior did not disturb Panther as he entered, for the oil-of-catnip scent was deeper than he had ever known it before. More had been used and it was inside a fairly closed system of limited air circulation. Panther stood in the twilight of the main room and mewed softly to himself. The scent drifted down to him from a high point in the smaller room beyond. Ignoring several mice that scurried out of his way and the hysterical anger of a land rattler that huddled beneath the floorboards, Panther moved forward to the door that led to the room beyond. The smaller room, like the larger one, and like the grounds that surrounded the cabin, was filthy. Litter and debris, human and animal, but largely human, were everywhere. Cobwebs breached every opening that an insect might use as an avenue. Stalagmitic animal droppings were hardened into place and, slowly surrendering their odors, had whitened with age. To any other creature the rooms whose threshold he spanned would have been rich in a variety of scents, most associated with decay, abandon and the passage of time. To Panther, however, the oil-of-catnip odor overwhelmed all others and after only a moment's hesitation and with a single lash of his tail, he left the floor. The tightening of the muscles in his hind legs was almost imperceptible but enough to drive the great cat up toward the shelf on the wall. He hit the shelf with his forefeet and before his hind legs could catch up, neither the cat nor the shelf was there.

Panther failed to right himself. He landed on his flank, with the shelf still between his forepaws. He was winded for a moment but not hurt, and he shook his head several times before bending his nose to the piece of lumber he now gripped. Sniffing deeply, he closed his eyes and let the scent roll inward and drug him. The euphoria that flowed over him masked out all reality and the scent was all there was.

When the shelf had torn loose from the wall, a quarter-inch nylon line had slipped off a three-inch common nail on the top of the plank, over which it had been looped. Lashing like a thin white snake, the line had whipped out through a hole in the roof and across the downward slope, dislodging several shingles and two irate herring gulls on the way. The rope seemed to have a life of its own and to be hell-bent on its own destiny. The weighted sliding door that had begun to slide downward as soon as the nylon noose slipped free of its anchor point hit bottom with a thud and effectively sealed the larger of the cabin's two rooms from the outside world. But Panther didn't know that. He was still sensuously involved with his displaced shelf. For the moment, at least, he was operating in a different space-time continuum.

The idea that cats never fall because of their own miscalculations or that they are never hurt in a fall is an old wives' tale. Cats do fall and although they are amazingly supple and breathtakingly agile, they can be hurt. It was no great surprise to Panther that the shelf had collapsed or that he had landed on the cabin's floor with an undignified thud. He had fallen before, he had misjudged, miscalculated and just plain missed.

The door made a very distinct sound as it hit, but Panther was preoccupied at the moment and didn't take any particular notice of it. Although he got up and moved around the smaller room once or twice, he spent almost the entire hour following his mishap romantically involved with the piece of scrap lumber. He mewed and purred and sniffed the wood and rubbed his cheeks against it with eyes half closed and ears laid back. It was as near to ecstasy as he could get. He was thoroughly engrossed for the full time and then small things began to slip into place and he stopped his purring, stopped his mewing, stopped caressing the wood. He stood up sharply and coughed softly once. Somehow he had been alerted.

Perhaps, even after an hour, he remembered the sound of the door falling into place. Perhaps he suddenly noticed the changed light value or perhaps he was able to detect that the flow of air within the cabin was not the same as he had known it to be on past mouse-hunting expeditions. Something was different, something had been altered, and by that altogether dumbfounding ability cats have of detecting the slightest change in familiar surroundings, Panther knew something was wrong. He coughed softly again and moved toward the door that separated the two rooms. His ears were half back and his lips lifted slightly in the corners. Panther was not only trapped, he knew he was. And like some of his kind, but only some, he was not resigned. He was ready to fight.

It was just past seven o'clock in the evening when Panther first realized his predicament. It was not until two hours later that he gave up trying to claw the door apart. Then he sat back to examine what his incredible exertions had wrought. For those two hours he had dug his claws into the wood and tugged, and had literally drummed his paws against the wood between raking movements. He had snarled and come close to shrieking in rage. No one can say if Panther fully understood his plight or what the measure of his fear might have been, but his anger was plain to see and hear. It could, in fact, be heard over a quarter of a mile away, where an old man sat hunched over in his small rowboat, nibbling toothlessly on half a sandwich. Doc Painter lifted his rheumy eyes and listened, contemplating the sound of the first panther he had ever owned. He decided he liked the sound and decided to wait until noon of the following day before offering the cat any water. By that time, the old man figured, he would have *"screamed hisself dry."* He would appreciate both the water and the old man better by then. Now that the old man had his cat, he had to break him, to force him to submerge his cat personality by any and all means possible, for that was the most important part of the exercise. The old man, who was known to all around as the greatest panther tracker the Glades had ever seen, would now walk in those places where he was known with a great male panther at his heels. It was the final indignity he would visit on the species of cat that had given him a reason for living. It would be the outcome of his last hunt. Such were his dreams that day.

Freedom of movement, freedom of choice, is the essence of being wild. A cat, any wild creature, exists as a reality only when it is free to follow every instinct, every whim, every stimulating combination of both that may come his way. A bird in a cage, however large and well kept, is not a bird in the sky. A pinioned swan in a pond that is little more than a manicured and siphoned bowl is not the same as a whistler or trumpeter overhead, and a cat in a cabin, trapped by wit and planks of wood, is not wholly a cat. It was an ironic and perhaps sad commentary that the old man who had lived in the wild and known the wildcats that had come before Panther should not have known this. He less than almost anyone else could be forgiven for his ignorance. He had come about as close to living like an animal himself as a man can come, and his morality was as much a cat's as a man's.

Like a fish that is pulled from the sea and begins losing its iridescence in seconds, Panther was a less luminous cat as he sat back on his haunches, rolled his head over to the side and spit halfheartedly at the door his claws had failed to devastate. He was no longer a wild thing but a thing driven wild by the foolish display of wit that had become so important to the old man. In the wild, Panther had served purposes far too profound for the old man to comprehend. As a prisoner in the cabin, he served none at all worth acknowledging.

When the old man appeared outside of the cabin exactly at noon the following day, Panther was crouched in a far corner of the smaller room. A dozen times he had tried the windows, tried to fit out between the solidly anchored two-by-fours and failed. He had tried the floor and found that he couldn't tear the boards free, although his claws dug satisfactorily into the aged and moisture-softened wood. He had tried the door again and had collapsed a three-legged table when he leaped upon it to try to claw down the ceiling. Without the table to help him he had leaped at the ceiling ten times, twenty times, a hundred times, in an effort to rip it apart and climb out and upward, to climb free of the room and its stench, free into the clear and relatively cool air that now lay so tantalizingly beyond. A dozen times he had hung from the rafters by his claws, had swung there briefly before dropping and spitting upwards, outwards at nothing in particular but at everything in general. He was frantic, furious and frightened. He was also thirsty.

Now, panting, with his ears laid back, the panther crouched in a corner of the small room and waited for the world to come to him with its next challenge, its new opportunities. He spit with unconcealed hatred as the old man's face appeared at the window between the slats of wood that made the cabin a prison. The old man peered in with no effort to hide his excitement. He whistled and whewed admiringly. He called out to the cat and cackled as his prisoner spit and snarled himself into a deeper fury. The old man was not ready for what followed, for he had killed so many cats so easily that he could not think of them as a menace. But for that instant at least, Panther was. Before the man could react, Panther was across the small room, hurling his full weight against the two-by-fours. The paw with claws extended came out between the prison bars faster than Doc Painter had realized any creature could move. It really didn't seem like a continuous movement. It seemed more like disjointed events. A cat in the corner and a cat at the window—with nothing in between!

The slashing claws barely missed the old man's face as he went over backwards. He didn't have time for so much as a single curse. He fell over eighteen inches off the up-ended box and was lucky to survive with all of his brittle bones intact. On the ground he *did* curse; he cursed and vowed to have his vengeance.

Slower now, because he ached and because he was not just a little bit frightened, the old man righted the box, picked up his rifle from where it leaned against the shack and climbed up to peer in again at the window. He leaned the barrel of his rifle on the sill to brace himself and screamed in at the cat, "*I could'a killed ya easy any time I wanted to. I could'a nailed yer hide whenever I wanted it. Don't act like that t'me, goddamn ya! I still can.*"

Panther lashed out again, passing close enough to the man to set him teetering a second time on his insecure perch. But this time he had the gun to brace himself with and he didn't fall. He was no less furious, though, and fired blindly into the room. The bullet tore past Panther's head a foot high and punched a ragged hole in the wall beyond.

The noise, though, didn't miss its mark and sent Panther flying away into the larger room, where he crouched low and hummed to himself. His ears were ringing from the monstrous boxing the sound of the exploding cartridge had given them. He was stunned and partially deafened. The old man stood outside and listened, wondering whether he had slain his cat in his anger, had in his haste shot a fish in a barrel.

He walked around the cabin and peered in through a peephole his sloppy carpentry of the previous week had provided and saw his cat crouching down in a far corner, shaking his head from time to time as if to rid himself of a troublesome mite. His trouble was not so easily disposed of, however. The explosion that had occurred inches from his ear had developed over three thousand foot-pounds of energy before the cat was even aware that an explosion had occurred. He had been slammed in his sensitive ears by a shock wave of staggering proportions. His nasal passages had been burned by the stink of the burning gases. His eyes had been temporarily affected as well.

The old man stared at the sickened cat, not at all sure of his own reactions or of what he wanted to do next. He had been subjected to several indignities, one of them being fear and another uncontrolled rage. He had suffered these at the disposition of the cat whose life he had, in his own eyes, saved by not killing it. He saw himself as the cat's benefactor. He saw himself as a kind soul who could have killed and had not, and who had been endangered, in a sense injured, and certainly insulted for his pains, for his efforts at humanity. To this man, this strange man full of years, each one of which had known death, no cat had the right to live, except by his leave. Cats were for killing and he, in having spared this one, performed an unaccountably large act of charity. How then, why then, should the cat act in this way?

Panther was not long in picking the old man up at the peephole. He saw the eye staring in and stared back, snarling softly. The old man with his killer instincts stared at the cat. The old man, not the cat, felt embarrassment. He didn't know what to say. He had addressed more words in the past half century to the carcasses of cats he had slain than to men he had known. He had to say something, now that he had a cat alive.

"*I'll be back, goddamn ya. I'll be back. Maybe timaarah ya'll want some water. It's gonna git hot in that cabin. Maybe timaarah ya'll be thirsty enough to come around. Actin' thata way to me . . .*"

♥ 9

Set Free by a Storm

From Panther!

There is no force in the places where animals live (unless, perhaps, at the bottom of the deeper sea) that has a greater power over animals than the weather. The weather determines what can grow and therefore what animals with what appetites can feed. Predators are governed by their prey and their prey by the fodder available.

Few animals are able to survive in almost any weather. Bactrian camels can; yaks, rats, and mice, some insects, and some microscopic forms can, but they are exceptional. Most animals survive within a relatively narrow band of temperature and humidity. "Harsh" is an often misused word and concept. Weather is harsh only to animals not suited to it. The Antarctic with temperatures a hundred or more degrees (Fahrenheit) below zero and with winds twice that of a hurricane in temperate climes is not harsh to penguins. We project when we say so. The desert is not harsh to the lizards, scorpions, and special snakes designed (read as evolved) for the desert, and again we think so only because we place ourselves in the animal's place and that isn't the best way to study natural history.

Weather, then, as much as any other single thing is a part of an animal's place. It would be impossible to understand how an animal evolved in a place or came to a place and stayed without seeing that animal in relationship to the weather that is shaping and honing the place itself and all the creatures it contains.

In an interesting way, I think, the weather is something else very special to an animal, any animal. Through the weather an animal is linked to the source of all power, the sun. The sun gives us our weather patterns. It is the engine that drives the tornado, the hurricane, and the peaceful, balmy breeze. Because it is a child of weather-given opportunities, the animal is also a child of the sun, as we are.

In the passages that follow, a weather system, by human estimation, goes mad. The story of Doc Painter is played out and the next episode in the life of the panther is made ready with the weather, a terrible storm, as the backdrop. The storm is a character, almost, in the way Doc and the panther are characters. They are linked together in a single embrace. In terms of drama the storm is also the deus ex machina, *the god of the machine of ancient theater. At the very last moment, when all mortal means have failed, the storm is lowered from the scenery in a basket to leap out and move events forward so the play can continue or be concluded.*

The first sign of radically dropping pressure within the huge tropical air mass had occurred over water, far from land. It had been nearly seventy-two hours before it was detected and reported. The radio message from the Algerian freighter had been somewhat vague as to position, but it was clear enough in substance. It reported to the National Hurricane Center in Miami that a tropical storm far to the south and east of Florida was shaping up and would bear watching. It was the kind of report that sets complicated technical wheels in motion.

Torrential rains moved ahead of the swirling mass and it became clear after four days that its erratic course had leveled out and that Puerto Rico was in its path. After two days of frantic activity Puerto Rico was ready. The storm hit and three people died, according to the news reports. Actually the number was eleven, but eight of them were nameless squatters, not the sort of people of whom either news or statistics are made.

The West Indies were next and the smashing force of the winds apparently increased. Anguilla was first, with four dead; Antigua was next, with five. Barbuda was hardest hit of all and thirteen known people lost their lives. A private plane out of Georgetown with four people aboard became the next casualty and the toll mounted. For almost a week Dora flipped back and forth, south and generally east of the Florida peninsula. Although it was clear that the storm was far too violent just to peter out, it still wasn't clear whether it would move north or head west across the Caribbean. There had been too many deaths already for anyone to expect sudden benevolence. This was a killer storm. Of course, it was possible that it could pass toward the north and east of the mainland. It was just that no one really expected it to.

Then, quite suddenly, the course was set. The storm began moving north, then northwest, then west. Its eye passed straight down "hurricane alley"—the ninety-mile-wide strip of sea between Florida's southern tip and Cuba—reversed itself, moved eastward, corrected to a northward track, struck between Marathon and Tavernier Key and whipped across the coast toward the heart of Florida.

The men who worry about storms were on top of Dora all the way. They knew her eye to be twenty miles across, her calm, low-pressure eye. Specially reinforced Super-G Constellations flew into it and circled there to keep track of the vicious lady's peregrinations. The men knew, too, that the storm reached up over thirty thousand feet and that the radius outward from the eye was over two hundred miles. The winds varied in strength but generally sustained their velocity at a hundred and fifteen miles an hour. Gusts, of course, reached much higher; up to a hundred and thirty miles had been recorded before the Weather Bureau mast was torn away and lost.

There were no weather fronts as such involved, for the storm had been born in a single tropical air mass. Since the eye was a low-pressure area, the direction of the winds had reversed at birth and were running counterclockwise as the storm moved toward the North American mainland.

Panther knew the storm was coming. All the animals of the Everglades did, at least all of those who lived a life exposed to air. And no one can say for certain that even those within the envelope of the Everglades water didn't know it as well, for the approach of the storm was attended by many changes, only a small portion of which were subtle. Here was something even a man could sense, which means that its symptoms were gross indeed.

As the storm moved closer, the barometric pressure dropped. This alone, in all probability, told most animals above the surface of the water that they were in probable jeopardy. There was a terrible stillness in the air and an ashy-gray quality to the sky, although no clouds were yet in view. Somehow clear blue had turned to gray and the ability of the water to mirror the world above changed, dulled, faded and then all but vanished.

The second clue to the changes that were about to occur was the sudden appearance of birds from the shore, a few at first, plainting their irritation, then scores, then hundreds, then a symphony of tens of thousands. Even the smallest hammocks that had harbored hundreds of birds now bore the weight of thousands as the flow inland increased. In places it seemed as if every branch, every reed, every solid object in view, bore its share of the winged burden. Telephone wires along Route 41 were lined as far as the eye could see and a dozen species joined the usual contingent of kingfishers in soldierly rows.

Windless, gray, with the pressure slipping downward, populated with new birds in uncounted numbers, the Everglades waited. It was the time of hush before the time of hell.

Even on a calm day there is *some* movement of air across the broad saw-grass flats. The sea is too close on all sides, the open land and inland waters too exposed to the Atlantic and the Gulf, for the air to be ever really still for long. But now the air was still. The incredible swirling air mass that was the hurricane seemed to be pushing a mourning silence on ahead of it, to prepare those about to die. Not a blade of grass moved, not a leaf trembled on its stem, and each bird sat rigidly still on its perch or moved only reluctantly when another came to rest nearby and unsettled those already there. The depth of the gray became more profound with every minute that passed. The temperature stood at one hundred and one degrees and the humidity lay across the world like a suffocating blanket of near-mist.

With the eye of the storm still two hundred and fifty miles away, Panther paced within his prison, more terrified now by the enclosure that held him than at any time since his capture. He tried the door again and then he stood on his hind legs for the thousandth time and tried to push his way through the improvised bars that blocked the window, the rough two-by-fours nailed in place by Doc Painter. When neither the door nor the windows would give, he retested the boards on the floor and threw himself upward against the ceiling. He hung there briefly, his claws hooked into the beams, and then dropped again to all fours. Looking up, he spat explosively in his frustration and then continued his pacing. He moved clockwise consistently and often brushed against the wall and the doorjamb as he passed from room to room. There were half a dozen places where his coat was rubbed bare along his left side. As the pressure dropped and the stillness that was a prelude to the ultimate explosion deepened, Panther's pace quickened. He was trapped at a time when only complete freedom could give him the barest chance of survival. Strangely, in his own way, he seemed to know it. His nervousness was so pronounced that it seemed as if he really did understand his plight.

The nurse checked each bed every hour. Unless there were other orders to interrupt her routine or unless a frantically pulled cord flashed a light over her desk and sent her hurrying down the hall, that was the attention the old and the dying received. There wasn't a patient in the ward for whom a place could be found in the intensive-care units; there were none for whom surgery was planned. The ward the silently moving nurse supervised was an exclusive club. It was for those for whom all had been done except the administration of a last-minute bit of comfort and, where possible, a last-minute portion of dignity. Since there is precious little dignity in death, in its immediacy or in the fear of it, there wasn't too much the nurse could do but try. Her professional life was a morbid treadmill. Each death blended in with all the others. The distinctions were harder to find, the demarcations less important with every day that passed.

The old man barely regained consciousness for a brief five minutes in the nurse's presence and she never had managed to get his name or any information about his non-existent family. Twice she had gone by his bed and stopped to listen while he whispered frantically about "th' cat," but she wasn't able to make anything of what he was saying, and moved on to other, more immediate duties. Delirium was all around her, every day, and she missed the significance of a hundred frantic messages entrusted to her for every one she got. A bedpan, a cry for water, or a patient violently protesting the steel rails beside the bed—these are far more immediate projects in a darkened room than the disjointed words that tumble forth on the shallow breath of a dying man.

Because it had been over an hour since she had seen him and because she fully expected to find him dead each time she stopped to check him, the nurse was doubly startled to find Doc Painter's bed empty. At first she stopped to think, to sort things out, for she wasn't at all sure whether the old man had already died and been removed. But no! She remembered it had been another old man several beds down the line. The old man with pneumonia, the old man whose card said *John Doe/"Doc" Age: Unk/Est 80*, the old man who was supposed to die, had come to within the space of the preceding sixty-five minutes and walked—or crawled—out of the ward.

Moving rapidly, the nurse checked all the obvious places. When she couldn't find him in the linen closet, in the ladies' or the gentlemen's toilet or on the stair landing, she called the front desk and reported the disappearance. Although the man's whereabouts was obviously a matter of life and death, it was fully a quarter of an hour before a search of the building and the grounds got under way. The life and the death of *John Doe/"Doc"* were no longer of great concern to society. Only last-minute comfort and his last-ditch dignity were really at stake, for the rest was considered fore-ordained.

The spinning air mass, still four hundred miles across, sent small fingers of high-velocity air out in jets as it whirled at nearly one hundred and twenty miles an hour, shifting and shuttling toward the northwest at fifteen miles an hour. The small jets shot across the coastline a good twenty miles ahead of the storm mass itself. The sea beneath the fingers was piled up, moving toward land in a kind of low wall, for the storm tide had formed. As the land underneath the tide edged up toward the beach, the mass of water built above it in a perverse counterratio. The blue that had turned to gray now turned to black. It was four o'clock in the afternoon and night had fallen, the storm night, the wind night, the night of death and destruction.

At the hospital the hunt for *John Doe/"Doc"* was called off. There were too many other things to be done, too many windows to shutter, too many patients to be moved away from windows. Beds were being wheeled into interior corridors, people were calling out instructions, and calling out in pain. The storm was minutes away and *John Doe/ "Doc"* was forgotten. One way or another, it had been decided, he would turn up. In any case, it would all be the same in the end.

The first air projectiles from the storm reached Bitterroot and stung into the trees with an audible pop. They had hissed violently as they shot across the water, leaving a trail of disturbance and microscopic destruction. Mirror-smooth water trying unsuccessfully to reflect back a gray world to a gray sky was suddenly marked by thin slices of wind. Their paths were arrow-straight, for they had been spun off the main air mass at high velocities. They shot across still waters and cut through hammocks like blades of air. Then came the noise, in the distance at first and then building, decibel by decibel, until it was a savage sound that filled the whole world.

The first few drops of water fell almost as soon as the first jets reached the hammock. It was a matter of seconds between those first drops and the torrent. It was blinding, staggering in its fury, and leaves and birds alike were stripped from trees and dashed to the earth and water below.

At the first sound of the wind, followed as it was almost immediately by the first sound of the torrent, Panther went nearly mad. He threw himself against the walls, against the door, against the window. He bled in several places where protruding nails had torn his hide and he was fairly screaming with rage and terror.

Glass windows along the oceanfront to the east that had been improperly prepared exploded outward as the whirling air dropped the pressure on their outer surfaces, turning the air momentarily trapped within the buildings into bombs. Merchandise from fashionable shops was strewn helter-skelter and bathing suits lodged ludicrously on cabbage palms while a royal palm further up Collins Avenue was draped in amber Canadian mink for a fleeting instant.

A policeman huddled in his car and repeated again and again through a loudspeaker that looters would be arrested. He kept on repeating this message, perhaps in the desperation of his fear, even though no one could stand on the street outside. In fact, no one was in sight. His message stopped when the horn was ripped from the roof of his car. Minutes later his car, with him still inside, began tumbling before the wind. He held his microphone in one hand and his drawn pistol in the other as his car tumbled over and off into the canal. He dropped both as he fought unsuccessfully to open the car door.

Glass flew everywhere and in less fashionable areas along the coast, overhead wires ripped loose. Communications were eroded, then finally brought to an end as aerials and broadcasting towers toppled and telephone lines disappeared under mountains of palm fronds. Then the sea in its piled-up storm tide moved onto the land and people could no longer recognize the streets on which they lived. Water cascaded everywhere and the smaller homes shuddered under the impact. A great many just gave up and collapsed and others were washed or blown away.

Doc Painter had had to stumble almost a quarter of a mile through the hospital parking lot and the undeveloped land beyond before he found a small dock with a dinghy tied to it. The boat didn't belong to anyone in particular, since it had been left behind several years before by an intern heading north for his residency. It was for anyone's use who desired it and was in fact occasionally pressed into service for short fishing trips by hospital personnel.

The first wind fingers were just knifing across the parking lot as Doc Painter stumbled forward into the boat. He was completely disoriented, totally contained within himself, as with automatic motions he untied the short, worn bowline. He stood and took the pole and pushed off as the jets of wind slashed in beside him on parallel courses. He stood there, poling his boat, heaven alone knows into which dream, a very old man dressed in a hospital shift. His legs and his back were bare except for the four neat little bows the nurse had tied behind him to hold the gown in place.

The old man's voyage was a hundred yards old when the rain struck. His knees fairly buckled under the impact from above. He turned slowly and stared up into the sky, screwing his eyes shut against the rain. His white hair streamed in the wind. He moved his lips and mumbled. The driving rain was pulling his hospital shift down over his shoulders. The top came undone, then the second bow, and then there was a small pile of white at the man's feet, standing out in sharp contrast to the gray world.

The radio tower on the hospital roof, the establishment's last contact with the outside, collapsed as the wind hit and seconds later the dock from which the old man had departed disappeared beneath the rising water that was moving up the river from the sea. The wind stung across the water, sounding like a thousand snakes. It reached the man almost instantly and as he turned, alone, quite naked and quite mad, he explained to anyone who would listen that he was going to Bitterroot to free his cat. His death was merciful. He neither felt it nor feared it because it happened to his body, which had been several worlds apart from his mind for nearly twenty-four hours. As his body and the boat in which it rode vanished before the wind and the water, the old man's soul moved off into its own eternal orbit and Doc Painter, the man who had slain two hundred panthers and captured one, ceased to exist.

. . .

The eye of the storm was destined to pass directly across Bitterroot Island from the southeast to the northwest. Winds in excess of a hundred miles an hour were to batter the hammock for nearly eight hours before the storm reached the west coast and spun out across the Gulf, heading for Louisiana and Texas with its promise of more death and destruction. Of the 1.7 tropical storms of dangerous intensity that have smashed into the Florida peninsula every year since the history of the area was first recorded, the one that headed directly for the hammock on which Panther was a prisoner was the third worst known. The barometric pressure dropped below 27.00 inches and held there for hours. At one point it edged down to 26.65, bringing it to within a hair's breadth of being the all-time recorded low for the Western Hemisphere.

Panther's terror mounted as the storm approached. When the actual winds themselves began hammering in across the flats, flinging debris up into great windrows and uprooting trees, the cat was driven into an almost insane fury. In a time of mortal danger, when his instincts were churning within him, nothing could have been more cruel than the severe restriction of movement. He crashed around within the cabin, repeatedly hurling himself against the walls, the windows and the door. Flying vegetation began piling up in the crude latticework of the windows, hung up on the improvised wooden bars, and wet branches with dead birds laced through them dangled within a paw's reach. But Panther wasn't after birds; he was after freedom.

The roof of the shack admitted water even in a light rain. As it was bombarded by the incredible torrents that poured from the sky before the storm and throughout its peak hours, the inside of the panther's prison was a sopping morass of filthy water. It poured in faster than it could drain out. At times Panther stood in pools that covered his paws. As the space beneath the cabin became saturated and the water began creeping across the entire hammock, the cabin stopped draining altogether and the water became hock-deep. This made the cat even more hysterical than he had been before. He bled from a dozen places as his repeated assaults on the nail-studded walls resulted in more and deeper lacerations and punctures.

He vocalized his misery by screaming against the wind, bellowing his full range of cat sounds. But the noise of the wind and the rain was so loud and the racket made by flying debris so awful that it is doubtful that an exploding cannon could have been heard, even if there had been someone to hear it.

As the wind reached its ultimate intensity, bizarre, seemingly impossible things began to happen. In Miami, to the east, a soda straw, an ordinary straw torn loose from a container in a fountain that was all but totally destroyed by the storm, whistled down the street like a projectile and imbedded itself in a palm tree to the depth of an inch and a half. A thirty-six-foot cabin cruiser traveled two miles inland on the insane tide and missed having a collision until it came to a one-story dwelling. It traveled completely through the building, emerging on the far side with a grand piano in its cockpit. There were two passengers on the boat before it hit the house, and three afterwards. No one was hurt.

After the storm, several dozen people would report having seen dogs and cats sailing by in the sky, and in fact they would not be exaggerating. Hundreds, thousands of domestic animals not recovered in time by families forced into shelters or ordered by law officers to evacuate lowland areas were lost in the few hours of hell. The number of wild animals lost cannot even begin to be estimated, but certainly it was in the millions. A twelve-foot alligator would later be found forty feet up in a tree. Somehow the tree survived the storm; the alligator didn't.

On Bitterroot hell was all there was. The entire island was flooded at least to some degree and 50 percent of all the trees over ten feet tall were lost in the first thirty minutes. Somehow, though, the ramshackle little cabin held. Then, from across the flats, one of the weirdest sights of the storm appeared. Several enormous royal palms that had been uprooted nearly ten miles away tumbled across the open space in a great curving arc, prisoners of an unbelievably powerful gust of wind. End over end they came, like mad acrobats in an insane circus. The giants somersaulted through the storm like living symbols of a world gone wild. Most of the tumbling palms missed the hammock, but one came directly ashore where the cabin stood and in a moment there was a pile of rubble where the small building had been.

Eventually Dora passed, leaving a shuddering world of devastation in her wake. Throughout almost an entire night she had screamed and wailed and tortured everything she could reach. Her destruction had been wanton, indiscriminate, without purpose or pattern. She had been a dealer in death, a killer, a living force gone berserk.

Dawn found the waters receding from Bitterroot. There had been times when the entire hammock had gone under, but the actual shape of the land had altered little. It did, however, look like an entirely different landscape. Very few tall trees were left and almost everything over thirty feet leaned precariously toward the west. The waters surrounding the island were littered with thousands of bird carcasses and the bodies of mammals, small and large, were everywhere. In places, five and six dead deer could be seen within an area of an acre or less. In the water, floating, beginning to rot from within and bloat, were birds and beasts, and on the land, wherever it was dry, thousands of fish. Along the shore of Bitterroot and every other hammock in Dora's path, great piles of trees and brush were stacked, packed so tightly that men could not have pulled them apart. Animal life, plant life and lifeless muck were piled and cemented together in monuments to Dora's strength. The number of dead animals that were laced throughout the windrows can hardly be imagined. Those that lived were usually the smallest and in the first still hours of dawn they began to creep out into the open, a mouse here, a frog there, a turtle and then a slender, glassy-eyed snake. They didn't hunt each other, not right away, for they were truly stunned.

It was only a matter of hours after the storm had passed before the rotten-egg stink of hydrogen sulfide could be detected. As the decay of newly dead plants and animals progressed, the intensity of swamp gas increased. In three days it was blinding in its strength. The sun beat down and what had been wet became dry, then brittle, then dangerous. Fires broke out as electric storms came and went, releasing far more voltage than water, and soon fires could be seen burning furiously in a dozen areas around Bitterroot. To swamp gas was added smoke, to devastation was added fire, to death was added death, and many of the creatures that had managed to survive the storm succumbed to this latest calamity.

It was into this world of death and evil odors that Panther had emerged at dawn. He had streaked like a bit of tawny lightning into a windrow after the tumbling royal palm had devastated his prison. As the walls had buckled out at the corners from the crushing weight on the roof and been gripped by the wind, Panther had sped through the first opening he saw and plunged panic-stricken into the first shelter he could find. In the ensuing hours his leg had been broken by another tree that added its weight to the pile and he had nearly drowned as water flowed deep where he lay before he could free himself. But he had survived where few large animals had. He was injured but alive, barely alive, and the same could be said for the land and most of what it contained.

The shattered right front leg would take time to heal and Panther hobbled, keeping it free of the ground as much as possible. Fortunately, he didn't have to hunt. Although panthers are not generally attracted by carrion, they will eat it when necessary. There was carrion enough to feed a thousand cats and he moved from deer to pig to bear to raccoon, nibbling a bit of the rotting flesh here, eating a little more there.

The torrential rains that had marked the storm had left the water level high, but in a matter of days it began to reestablish a more normal level. Then the world of the Everglades began to rebuild itself, to seek again an economy of life, a balance and a working system of interrelationships. An ecosystem, like a rubber band, has a memory. No matter how it is stretched, it will try to re-form itself again.

The first wildlife to return were the birds. Thousands had escaped to the north and some to the more risky south, and now they began filtering back to reestablish their roosting sites and their rookeries. It was not many weeks after the storm had passed when the first fall migrants from the north appeared and in the waters there were fish again. Insect life had survived without a dent, for a billion bugs here or there really does not matter.

One day a raccoon wandered ashore on Bitterroot, heaven alone knows from where, and a few deer appeared the next evening. There were frogs and there were reptiles, and soon, although it still smelled of death and decay, Bitterroot began to witness the reestablishment of its animal community. Plants and trees began sending up new shoots and leaves appeared on branches that had been stripped. Through it all, as his leg slowly knitted and his other wounds healed, Panther moved slowly, cautiously seeking again a new rhythm for a life that had been disrupted and all but destroyed, first by an old man's folly, then by the hurricane.

Six weeks after the storm, after six weeks of knitting and the careful husbanding of strength, Panther killed a deer. He felt an agony of pain roar upward into his shoulder as the frantically bucking doe tried to tug free, bringing unbearable pressure to his recently injured leg, but he held on and ended the animal's life by seeking the base of her skull with his stabbing teeth. He drank her blood first, then he opened the animal's paunch and ate the viscera. He scraped a few leaves together and observed the careless ritual of covering the cache. He returned to it the next day, and the next, and then, after eating off the deer for the fourth time, Panther left Bitterroot Island, left the hammock he had claimed as his own long ago and headed into a new life. Once his island had meant security, stability, a safe place to be, a safe place to return to. But Doc Painter and Dora had conspired to change all that and his hammock could never again be the same. Now that his leg was healed, although there would be pain for several months when he was careless in its use, he was free to go, free to wander until he found another place that was without memories of sick old men and savage storms. He was once again the essence of life, a wild thing free to follow his wild ways. It had been ordained, long ago when the cat was designed, that its life should be one of danger and death. That is the way the cat was built and that is the way he was fitted into the world that was there before him. Without a job to do, without a niche to fill, the cat would have been surplus and would have been destroyed in a world where plans are not so casually made. A wild thing in a wild place has a wild life to lead as well as a job to do. Panther was such a wild thing and the Everglades is still such a place in part. The job and the life still lay ahead of him and Bitterroot of the evil memories, by his own wild choice, was to be no part of either. Panther, it must be remembered, had still not reproduced himself in kind and therefore had not fulfilled his original purpose in life.

● ● ●

There was no special reason, at least none that a human being could discern, why Panther chose to move toward the northeast once he left Bitterroot, but that was the direction he selected. It was shortly after dawn when he stepped off into the shallow water and headed across the open flats toward a smaller hammock several miles away. He moved along at a steady pace as the hammock grew larger and larger before him. There were closer ones off to his left and his right, but he had selected his target and kept himself on course toward it.

When he was about a hundred and fifty yards short of the little hammock, a thudding roar split the world apart and an airboat came skidding around the north end of the island, ploughing up a ridge of water before it. As he hit the curve of the island at fifty miles an hour, the hired boat's driver had kicked his rudder hard and amid the shrieks of his three passengers the boat skidded sideways for over a hundred feet. It was what the passengers were paying for. The wave set in motion by the boat hit Panther where he stood frozen to the spot. The wild skidding and quick recovery of the speedy, agile craft brought it to within a hundred feet of Panther and for a moment he didn't know what to do. Then as the driver cut back on the throttle and the roar softened slightly, Panther broke into a lope, fleeing toward the small hammock's shore nearby.

One of the passengers on the boat saw Panther first, even though his seat was lower than the driver's. As the boat rocked to a stable position and the engine was cut back to an idle, the three passengers and the driver all stood and called out to each other as Panther ran in one splashing bound after another until he was lost in the heavy growth that came down to the water's edge. At a little less than half speed, the driver worked his airboat around the hammock until he determined that the cat would lay up there for at least a day or two. *"We scairt him good,"* he mused aloud. *"He sure was scairt pretty good."* The driver of the airboat was sixteen years old.

❧ 10
Suspended in Time

From
Source of the Thunder

It is not unusual for natural history writers to be so taken up by some element of an animal's place or some aspect of the animal itself that a larger than usual emphasis emerges when they try to capture the essence of the animal and its place in words.

I had gone to watch the California condors a number of times and one word kept coming to mind, ancient! Indeed, they seem prehistoric and, in fact, are. Whether wheeling overhead, sitting in a tree, or sunning themselves with wings half extended and shoulders hunched as they turn around and around high on a ledge above the valleys where they seek carrion, the condors seem to be animals of another age.

Add to that fact the relatively unexceptional nature of the sparsely wooded hills where they were until recently just about holding on for those last few moments of their long history, that area north of Los Angeles, and you have the irony. The most interesting thing about the condors' place, or perhaps the condors' most interesting place, is not geographical but temporal, their place in time. I saw no other way to start my story than to establish

*that fact. The condor, then, in its place in the history of its planet,
and ours:*

Several billion years had passed since the sun's middle-sized child had established an orbit around its parent star. It was one of nine, larger than four, smaller than four, but it was unique within its own family in a number of ways. Its position between sister planets Venus and Mars placed it approximately ninety-three million miles from the sun. That distance assured the infant of a temperature neither too hot nor too cold to permit certain critical chemical reactions to occur. New amalgamations of basic elements would be formed. One day, some of these substances would reproduce themselves; they would live.

The mass of the new planet, and a density greater than that of any other in the family, had provided it with a surface gravity strong enough to hold a thin layer of gas to itself. It would have an atmosphere. That which would live would be sustained.

Lifeless, turbulent, fierce in the extremes of its infancy, the new planet spun around on its own axis once every twenty-four hours while hurtling through space at a speed of eighteen and a half miles a second. The matrix of space being close to a complete vacuum, even these terrific velocities failed to abrade the new child of the solar system. Tightly gripped by the sun's gravitational field it spun, showered with debris of both matter and energy, building its own atmosphere as it went. Gases boiled free from the yet unestablished rocks and the shape and texture of the ball evolved from internal forces and the basic laws governing matter in the cosmos. Cosmic radiation rained down unimpeded and chemical reactions occurred that can barely be contemplated. Although much of what was really important happened on a microscopic level, the larger events must have been horrific in force and violence. The din, unwitnessed though it was, must have been incredible once there were enough molecules in the gaseous envelope to bounce against each other.

And so the first eons passed. One planet out of nine moving around a sun that itself was part of a system of a hundred and fifty billion suns (the system being one of billions upon billions of systems), surging through its elemental times creating its own history. And that history included the most awesome of all miracles, the evolution of raw chemicals into a self-duplicating state called life. That state, once it had come into being, assured the planet of immortality, for mortality and immortality are a continuum, one but an extension of the other.

After the planet had been some three billion years in orbit, blue-green algae appeared and the complex molecules began their march. Upward from the slime and the reeking chemical baths cradled in rocks still hot from their early history, life spread and diversified. The forms life took became increasingly more complex. Single cells learned how to live together and, in time, strange creatures, some even monstrous, inhabited the planet. Scorpionlike eurypterids more than eight feet long, brachiopods and trilobites, clams and cystoids, horseshoe crabs, shrimp, snails, and jawless fish spread, changed, spawned new and more successful forms, and vanished. A few resisted time and have lasted on to our present day. Ammonites and nautiloids, lungfish and sharks, labyrinthodonts and forms we do not yet know moved onward. It was a time of experimentation and the diversification was endless.

Competition became more keen and the seas were not enough. In time animals were crawling and walking upon the land. Some forms left only a trail of slime to mark their passage, others footprints. Those that were compelled by their biology to return periodically to the sea gave rise to higher forms that were not, and reptiles, bizarre, enormous, diverse reptiles arisen from amphibians, themselves arisen from fish, ruled the world. Some of these, awesome beyond belief, returned to the sea and prevailed there as their cousins reigned on land, where mountains were yet to be born. It was the time of giants and although the brain complex enough to conjure up nightmare images was still hundreds of millions of years in the future, the nightmarish creatures themselves were a reality. Dragonflies were as big as modern hawks, salamanders so large their bones would one day be confused with those of men.

Once upon the land the reptiles were free to explore and experiment with many ways of life. No avenues were untried and those that elected to remain behind perished for the most part, while those that went on ahead to new ways ended in glory or vanished. Glory, of course, was survival. Five large suborders of reptiles spread out with thousands of species. One of these suborders, the thecodont reptiles known to us now as Pseudosuchia, had an appointment with destiny on two levels. From their midst arose the ponderous dinosaurs who for millions of years would possess the planet and then forever tantalize later creatures who would come to know their bones. Even earlier they had evolved specialized forms that took the thin layers of gas above the earth into their ken. They learned to fly.

The archosaurs, giant reptiles still, solved the problem of flight with staggeringly large batlike wing-membranes as much as twenty-five feet across. Although they flew, they were not birds. Their jaws were heavily armed with the gripping and tearing teeth of the carnivores and only the earliest suggestion of feathers could be detected in their epidermal scales. The hint existed only in that the scales were twice as long as wide with fine striations running out from a central axis. Still, the hint and the promise were there.

The earth, although now stable in its position within the solar system, had not yet found inner peace. Volcanic eruptions tore the surface apart and massive quakes shook whole continents, sending frenzied seas across the land. It is believed by many that entire continents shifted and shunted around the surface of the globe during those terrible times. Cracks and fissures opened, and burning gases spewed forth in geysers that painted the sky. Steam rolled across the land in billowing clouds and the mists of time concealed millions of secrets, any one of which would be critical to our understanding of our world if it could be revealed today. But, too often, what went into the mists dissolved there and today we are far richer in conjecture than in facts. In our souls we can dream of those times, but we do not know them.

One of the most crucial of these secrets may be lost to us forever; more accurately, it was a linked series of happenings. We have no fossil record of the many stages those primitive flying reptiles knew until the relatively recent period of one hundred and fifty million years ago. Then it was that *Archaeopteryx* flew, during the Jurassic period. Its flight was not strong and probably carried it for short distances only. It was a light bird, crow-sized, although its bones were not yet hollow. The bones of its tail were still unfused and had twenty free vertebrae to accent its reptilian beginnings. A single feather grew out from each side of these unbirdlike tail vertebrae. There are no links to connect this first known bird with the flying lizards, but from somewhere within the reptilian scheme they had come.

There were many other ways, too, in which *Archaeopteryx* differed from the birds of our time. There was a distinct hand with free-moving and well-clawed fingers, the pelvic bones were unfused, still held together by ligaments after the fashion of the reptile. The teeth again linked it to its reptilian past. Yet, it was a bird and it flew and it had within it the potential of the eight thousand species of birds we know today. This wonderful creature was without beauty but even in the impressions of its bones we can detect its solemn vow to one day grace the world with color, song, and charm.

Once again the veil of time floated across a world no longer infant but still new and rich in possibilities. The birds that arose from *Archaeopteryx* were lightweight and fragile and left fewer records than heavier-boned creatures on other evolutionary tracks. As we seek its path today, this branch of evolution is far more tantalizing than revealing. It skips in and out of fogbanks, revealing itself only rarely, and then briefly, giving us only the barest hints of the wonders that once were and can never be again, at least on this planet. The fact that we can never know their colors, their songs, or their nests must forever taunt us. What trills and runs did dinosaurs hear each spring as these ancient birds sought their mates? Or were they mute, or did they hiss like snakes?

Change was constant, we know, and hundreds, then thousands, of new bird species evolved. Each had packed within it signs of its reptilian heritage and promises of its avian future. Thirty million more years passed, with the force of evolution surging into every new channel it could discover, and the Cretaceous shales were laid down upon the earth. In that time, the modern bird form was suggested. Gull-like birds flew over land that would one day be known as Kansas, Montana, and Texas. Their brains were small compared to those of birds today but the reptile phase was demonstrably over. Elsewhere, loonlike birds, still equipped with teeth, fed on fish, and birds not unlike our herons, geese, and cormorants found their places. The dinosaurs were already dying off, for their potential had been exhausted and the very qualities that made them awesome proved to be their undoing. In the terrestrial environment their size was a cul-de-sac. In the seas, the mammals had better evolutionary ideas.

Some sixty-five million years ago, when most mammals were still small and harmless, twenty-seven families of modern birds already existed. Grouse, swifts, auks and penguins, sandpipers, bustards, grebes, and cuckoos. Evolution had already established these paths and saw in them systems worth developing. She began their refinement.

Within another thirty million years Nature had found the ways of the storks, the plovers, the turkeys, pigeons, parrots, and sparrows to her liking. Ten million more years, the petrels, falcons, and oystercatchers added their survival skills and the Miocene was born. Land areas that no longer exist were populated and seabirds flew over open oceans where island chains were yet to be born. The only thing in which Nature sought permanence was in the force of life itself.

All of this progress was not without its casualties. Nature was quick to discard experiments whose promise was not as great as that of others and the extinction of species was as ready as their evolution. It was in the Pliocene, no more than ten million years ago, that bird species probably reached the maximum. By that time all modern bird genera were in existence, and many we no longer know. Mammals, also descended from reptiles, were evolving as rapidly and the hint of man was already in the loins of ancestral apelike creatures. It may have been that long ago when the first stick was raised up as a club. It may one day be demonstrated that this was the one irremediable mistake Nature ever made. By creating a creature over whose destiny she might one day lose control she may have sacrificed her hold on this one planet, at least.

And so the form of life we know of as birds had come to be. On the gases that seeped up through the rock crevasses of continents still resting on molten beds they flew. They were hatched by the billions, died by the billions, but the power of their flight into time never weakened. It was a successful experiment and from the moment it was launched its outcome was assured. No catastrophe could stop the force of bird life, no competition could do more than sharpen it and force it to improve itself. Time was the honing stone, Evolution the edge-hungry blade. It is only in the last few moments of bird history that any creature has lived capable of the concept of beauty, but that beauty has been here these millions of years. It was Nature's plan then and now, we can believe, that it should always be. There are too few flaws in the design to allow us to think otherwise.

For millions of years the Cretaceous seas had been depositing their sediments in the depression that ran from far north in Canada to deep into Mexico. Between eighty and ninety million years ago, while the dinosaurs and toothed birds still held sway, a revolution began along the line of this enormous trough. The Laramide Revolution, one of the most awesome upheavals in the long and tortured history of the Western Hemisphere, began the thrusting and faulting that would eventually fold a portion of the entire continent upwards until the ragged edges of rocks torn from their bed beneath the sea would reach the sky. The Rocky Mountains were born. From Alaska to the heart of Central America the colossal peaks rose in a range five hundred miles wide in places. Not since pre-Cambrian times, nearly five hundred million years earlier, had such gross physical forces played upon the crust of the earth. In the east, two thousand miles away, the Appalachian Mountains were already a hundred million years old when this new range began its march from below sea level to the sky.

Although erosion began immediately, and some peaks were already beginning their infinitely slow decline before others arose nearby, the great chain of upheaved rock formed then, as now, an enormous barrier. On one side, in the east, the slopes descended to monotonous flat plains. On the other, in the west, they ran down into untidy land that again and again flung great parapets upward until at last the continent came to rest in the sea.

It is likely that active volcanoes throughout the region added their influence to the geological confusion and over a period of millions of years the continent struggled to stabilize itself. Thrust after thrust contributed to the complexity of the revolution until slowly the shape of the land was established, and it rested in relative quiet for the millions of years the forces of heat and cold, wind and water would need to carve the monoliths back down to the level of the sea. That period is still in its infancy.

No one can know how many animals and birds, individuals, species, and perhaps genera were lost in the upheavals. How successful the ancient birds were at escaping the cataclysms we cannot know but through and beyond it all, bird life on either side of the mountains continued to develop. As the mountains came to rest, first in one area and then another, the winged creatures, and then those that crawled and walked, began a reinvasion and the Rocky Mountains, too, had their fauna.

Although the biography of every living creature properly begins when the planet was still an infant, the story of our condor will be reckoned from the wonder-time of North American fauna, the Pleistocene, the final epoch of the seventy-million-year-long Cenozoic era. Seldom has animal life known more fascinating diversity or greater numbers than during this epoch. Men living today who have an interest in such things can but weep for not having seen it. We can know it now only from the mountains of bones we uncover. They taunt us, they tantalize, and we struggle suspended between awe and frustration. What a time it was!

By the time the Pleistocene dawned, the horse, a line of evolution begun in North America seventy million years earlier, had established itself in relatively modern form and had begun its march across the land bridge that existed on the roof of the Pacific Ocean to spread throughout Asia and finally Europe. Although extinct here by the time settlers arrived from Europe, they existed then in herds the size of which we can barely imagine. Stocky, probably uniform in color, heavy-headed and bristle-maned, they grazed their way over enormous plains and scattered in wild confusion at the approach of carnivores.

Giant wolves, whose counterparts exist nowhere on earth today, hunted, probably in family groups and occasionally in larger packs. *Smilodon*, the saber-toothed cat, stalked his victims as a solitary hunter and literally stabbed his prey to death with canines that were flattened and scimitarlike, with finely serrated edges. Biting, in the normal sense of the word, was impossible, for these enormous tusks blocked entrance into the mouth. The cats probably struck much as venomous snakes do today. They ranged from the shores of the Pacific Ocean eastward to Pennsylvania, and existed in surprising numbers.

That such great numbers of these terrifying cats and their canine counterparts, the dire wolves, thrived, is further testimony to the numbers of prey animals that were there to feed them.

The giant ground sloths, short-faced immigrants from Central and South America, spread across the land to fall prey to the marauding bands of wolves and the solitary stealth of the stabbing cats. When on all fours, these sloths stood more than four feet at the shoulder; they have no counterparts surviving into modern times. They were found everywhere, inoffensive browsers with neither speed nor sudden wit at their disposal.

Pronghorns, erroneously called antelopes, similar in most ways to those few that still exist among us, ranged in large herds and frustrated the hunters with their speed and wariness.

Large camels wandered across North America; mastodons, no less elephantlike than the mammoths, existed in vast numbers. Short-faced bears even larger than the existing giants on Kodiak Island, coyotes related to those of today, and many smaller creatures filled each available niche in the ancient ecology. Peccaries, lionlike cats distinct from both the contemporary sabertooths and pumas, tapirs, foxes, and many other forms that would not seem strange to us today were there. Rattlesnakes were by then an old idea and lizards, diminutive reminders of far more ancient times, abounded.

The wonder of the Pleistocene was not limited to animals that wore fur and scales, however, for the birds were as wondrous, and here our story starts.

．　　●　　．

The summer, like the thousands that had preceded it, had been relatively dry. Although the sea was only a few miles away to the west, the sandy flat with the scattered oak, hackberry, and juniper stands was like an internal zone. The low mountains that cupped the area in on three sides were not very impressive but rose just high enough to give the winds sweeping in from the sea a vertical lift that kept the skies over the bowl fairly clear much of the time. On days when the winds were slow or absent, moisture-laden air drifting in casually from the warm seas lingered there and the humidity would rise. But active precipitation was light and scarce.

Although ringed by low hills, the sandy flats were not isolated. Numerous cuts through the hills, many of them nothing more than reminders that rivers, like animals, live and die, provided pathways for wildlife to and from other regions. The rising thermals were particularly beneficial to the soaring birds and they came and went, often at high altitudes.

It was the late Pleistocene, and far to the north and east the third interglacial, a minor one compared to the one that had gone before the now defunct Illinoian glaciation, was coming to a close. Although relatively short in duration it had sent the ice sheets creeping guiltily back toward the Arctic and had seen the temperatures in America's midlands rise sharply. It had guided the northward migration of millions of animals whose ancestral forms had been driven southward thousands of years earlier. This was about to change again, however, as for the fourth time ice sheets were forming and moving toward the south; the Wisconsin glaciation was on the march.

Further to the north, in the flatlands of central Canada, forests were already vanishing, turning to pulp beneath the grinding weight of ice a mile or more deep. Across an enormous front, the ice sheet was edging forward as inexorably as the movement of the planet. The frigid air mass that hovered along the front sucked moisture from the air and deposited it in rime and frozen frost. Snowstorms swept across the sheet, adding to its thickness. At a depth of twenty-eight feet the snow compacted itself into ice. Electric storms crashed along the glacier's face and once again the continent trembled before the fury of an ice age.

To the south and west, though, matters were less urgent. The land was changing there as well, but in a more orderly manner. The higher mountains had their own glaciers but they were small, personal affairs. Snow fell and occasionally a valley would be lost for a few months, simply filled up with snow, and rockslides answered snowslides, but all this was seasonal and local. The mountains grew smaller each hour of each day and the alluvial fans grew broader and thicker. Even if there had been witnesses, these events would hardly have been discerned. Their timetable was protracted and little happened day by day that could be seen.

Beneath the surface of the earth, though, in that sandy area with the juniper stands just in from the sea, other forces were at work and these could have been witnessed had there been men to do so. They were events that cost, in time, tens of thousands of individual creatures their lives.

Beneath the sandy flats were folded layers of variously colored clays. A yellow clay lay on top, lightly sprinkled with recent debris, below that brown clay; in some areas there were thick streaks of blue. Veined through it all were layers of oil-stained sand, buried pools of liquid tar, and pockets of gas. Seams of bituminous material threaded through, and black clay, part petroleum and part inorganic sands, rose and fell within the matrix according to the chances of fold and distortion. It was uncertain land, an area to be known as the Los Angeles basin in a distant future.

One quiet day, when the winds from the sea barely stirred the dust devils that lay sleeping in the sand, when the sun was hot and yellow, a distant grumbling could be heard. It was a discontented sound, a complaining one, as if a mighty force were angry at the obstacles it found in its path. Suddenly, with no more warning than that, a thick stand of junipers seemed to rise straight into the air and then part. For a brief instant the ground that rose with them, still carrying their roots, looked like a round-domed hill, but then it disintegrated as the huge pocket of gas that had been forming seventy feet down exploded upward. The trees scattered in all directions and where flatland and then a fleeting dome of earth had stood there was a hole, nothing more. The sound of the explosion was muffled and there was no fire. The hissing died, the gases joined the atmosphere, and once again internal forces had changed the landscape, although in this case the change was small and of local influence only. Still, a cavity in the earth is a dynamic thing and many forces are summoned into play by its creation.

Within minutes after the explosion occurred, the land began to bleed. The upward thrust of the overburdened gas pocket had slashed through seams of oil-bearing sand and had ruptured several trapped reservoirs of tar. A thick, black liquid began seeping down the walls of the newly excavated hole and pools formed in the bottom. The winds rose and thin layers of sand rained down only to be absorbed in the pools and then covered with more liquid as the flow increased. Like an osmotic membrane, the walls of the hole drew liquid from the surrounding strata. Quartzite pebbles and igneous boulders tumbled into the growing pools and in places small landslides occurred in the coarse sands. The pool filled, hour by hour, day by day. As veins emptied themselves and their open ends were smothered by the deepening tar, gas followed the channels and surged in below the surface. Worming its way upward, the gas bubbled at the surface in small, tarry volcanoes. Gas burping free into the air knocked low-flying insects down to struggle briefly on the clinging tar swamp. Small rodents exploring along the edges of the pit found the footing uncertain and each day some were caught in collapsing ridges and swept down into the pool. Their struggles were hardly more impressive and no more fruitful than those of the dragonflies and mosquitoes. The fossil treasure of the La Brea tar pits had begun to form.

It did not take many months for the pool to fill, and then overflow. The seeping liquid found open holes, places around tree roots, animal burrows, and other gas evacuations and flowed in to fill these as well. The area was dotted with black splotches but the land was still essentially sandy. Often high winds would scatter sand across the flat and the exposed tar would be covered, made briefly invisible by the crystalline debris. Eventually, though, the grains of sand would pock the surface of the tar with their own weight and then vanish, sinking slowly down to mix with the growing collection of bones that was accumulating near the bottom. For days and sometimes weeks on end the tar would be shiny black again, marked here and there with a newly exploded bubble of gas, until new winds and new wind-borne debris littered the area.

The cow camel stood seven feet tall where her back arched slightly above her rib cage. The top of her head was more than eight feet from the ground as she ambled along, her head thrust forward. As much a llama, really, as a camel, she did not have the pronounced humps of her Asiatic cousins but she was unmistakably of their cut. Her calf stayed close to her, for the survival demands of their species had instilled in them an instinctive alertness, a natural caution. Proof of the predator lies in the movements of the prey.

As they moved out into the open they passed through a loose stand of live oak, then some manzanita. Far off to their right a herd of enormous bison, as high in the hump as the top of the camel's skull, grazed slowly in the late afternoon quiet. Further on, two imperial mammoths, nearly thirteen feet in height, shuffled forward on their pillarlike legs, their heads seemingly weighted down by their enormous incurving tusks.

The powerful sabertooth, about the size of an African lion today, had been resting up in a hackberry grove during the heat of the afternoon. Truculent, small-brained, and aggressive it had been alone most of its life. It did not get along well with others of its kind, or with any other animals for that matter. It attacked what it met and both inflicted and received crippling injuries in many of these encounters.

Now, in the late hours of the sun, it sensed the cow camel and her calf. It rolled over on its side, then onto its belly and watched, head low to the ground. Off in the distance a herd of small pronghorns stood at alert, the herd buck barely two and a half feet tall. They were too swift, though, and the day still too hot. Even if he revealed himself, he knew instinctively, the camel's flight would be hindered by her concern for her calf. His choice was simple, obvious.

Slowly inching forward with his weak hind legs and powerful forelegs folded under him, the sabertooth worked himself to the edge of the hackberry stand. Then he stood and in a second was in open pursuit. He could at least have the calf. The cow camel spotted him immediately and, calling her alarm note, galloped across the flat with her calf bleating piteously in pursuit. It appeared for an instant as if they might outdistance the cat but suddenly the camel's front legs buckled and she plunged forward, then over in an awkward somersault. Her front feet were mired and both legs snapped as she went over. The calf, because of its lighter weight, was able to take two full strides onto the surface of the tar before becoming entrapped. The cow's struggles enabled her to get her head free from the mire but she was doomed, unable to right herself.

The sabertooth was only a few strides behind and in a final leap landed on the stricken camel. His weaker lower jaw swung clear and again and again he stabbed and twisted his incredibly elongated upper canines into the camel's belly and chest. His strokes were powerful enough to splinter bone and the camel's ribs collapsed. The cat's nostrils were far up on his snout, conveniently back and out of the way, and he was able to push his muzzle deep into the gaping wounds his teeth made without having his breathing hindered. The free-flowing blood coursed along corrugations in his upper gums and he sucked and swallowed, sucked and swallowed. He then began to eat.

The calf had died within minutes after the sabertooth had killed its mother, its bleating having summoned a number of spectators to the side of the pool. During the few minutes of furious activity that occurred immediately after the camel became mired, a pocket mouse edged near the pool's outermost seeping and was struck by a rattlesnake. After ingesting its prey the snake slithered across some apparently firm sand and sank out of sight in the tar. A two-foot-tall *parapavo* turkey became entrapped at about the same time in another pool nearby and increased the rate at which it sank by its frantic struggles. When only its head remained above the tar a raven landed on some sand nearby and began pecking at the turkey's eyes. When it tried to lift off, the raven found that it, too, was caught and soon vanished beneath the surface.

The sabertooth's perch on the cow camel's belly kept he clear of the tar during the time he fed. When he was satisfied, though, and turned to leave, he found his hind legs gripped by the viscous trap. Struggling violently, his numb brain barely able to grasp the nature of his plight, he finally pulled himself free. He then turned and stepped off the carcass as if nothing had happened. His forelegs sank into the tar up to the shoulders and the recently fed saber-tooth cat was doomed to die the death of thousands in the greedy La Brea tar pools.

A thousand feet up, eleven incredible *teratornis* circled on their twelve-foot wings. Undoubtedly the largest flying birds the world has ever known, these forty-pound scaven-gers waited patiently for movement to stop. As the saber-tooth's last convulsive kicks ran down his hindquarters (his head was submerged and he had suffocated), the *teratornis* began their descent. They spiraled down until one after another they landed on the exposed portions of the camel, her calf, and the great cat. Before they were ready to lift away, three of them would be caught and their bodies would be gone before the next sunrise. None of the ani-mals in the area seemed to be able to learn the danger the black pools posed and species after species contributed carcasses to the bone collection of the pits.

Just slightly lower on the carrion-eater's pecking order was another great soaring bird, the California condor. Specimens of this species, too, had seen the flight of the camel and her calf, their entrapment, and the sabertooth's attack. While circling high above they had watched the *teratornis* land, feed, and depart leaving three of their kind struggling in the tar. When the remaining eight giants had departed, the condors started their slow, circling glide down. Cautiously one, then another, then two more landed on what remained of the carcasses. All four fed on the still ample remains, one concentrating on a *teratornis* before it was quite dead, and three finally managed to lift away. Only one became trapped. Of the three that managed to get away, one separated from the other two and flew off toward the north. Encountering a strong updraft, it gained in altitude and slanted toward the northeast. Within an hour it was flying over a deeply cut valley with high, predator-proof rock walls, and then it began its descent. It came to rest on a bare oak that grew out of a hundred-foot-wide grass-covered ledge a thousand feet up from the canyon floor. It remained on its naked perch for ten minutes, scanning the sky above and the canyon below. Satisfied with its solitude, secure in the belief that it could proceed without disturbance, it spread its wings, caught the wind and kicked free. It circled the canyon twice and came down on a rocky ledge several hundred feet below the canyon's rim. From inside the rocky cave a small downy ball came forward hissing and flapping its stubby wings. Within moments the heads of the two birds were locked together and the adult regurgitated a chunk of sabertooth tiger flesh. The infant condor withdrew its head from its parent's bill and began to feed. With its peculiar shuffling gait the parent bird moved out of the cave to the rocky ledge overlooking the canyon and stared off into space. Its bright red eyes had a timelessness in them. As windows to its brain they somehow belonged as much to the future and the past as to that particular present.

As it watched, a form drifted in at the end of the valley. The mate circled slowly until it found a convenient draft and then drifted straight down toward the cave. When it was a hundred yards off, the male lifted free of the ledge, fell for a moment and then caught the rising air. Together the two birds rose up, up beyond the rim of the canyon, up until they were twelve thousand feet above their valley. They selected a flyway, a route already thousands of years old to condor and condor ancestral forms and flew toward the south. A roughening in the air told them a storm lay in their path and they climbed to fifteen thousand feet to avoid it. At that altitude they encountered thick clouds. Unable to climb over them, without a chance of flying under them since they reached the ground in the heavy storm weathering the area, the birds curved toward the east, cutting across the front with its potentially dangerous winds. In a valley forty miles to the east they found a quiet area and came to roost. In thirty hours the area would clear and they would return to their own valley carrying more food for their single chick.

II

The Condor in His Unexceptional World

From
Source of the Thunder

Sooner or later, though, the nature writer has to get to his or her leading character. The condor has to be lifted bodily from its ancient setting, its place, and put forward in time so the drama of a single bird can begin. That is what, hopefully, a reader can be made to care about.

There are many ways to handle any problem in writing, to aim at least for any goal. I could have gone cold turkey (no pun intended) into an egg and been hatched with the baby as I have in other books with other animals, or I could come forward in time in a relatively few pages and emerge onto a modern stage, plateau, to play out the recent drama. In this case I opted for the latter approach. It seemed more in keeping with the idea that the condor's place is most profitably viewed as a place and movement in time than geography. Everything a writer does, every word in fact, is a judgment call. And any writer might elect to do things in a different way.

The California condor that flew then, in the time of the mastodon, ground sloth and sabertooth cat, was not unlike the California condor we know today. It was, almost certainly, an ancestral form only slightly larger in size. We *assume* it looked very much like the bird we now know, for so its skeleton would have us believe. There was a major difference, however; it was neither rare nor restricted in range. The pair that carried food back to the chick in the cave when the storm had moved off could have flown fifteen hundred miles to the north, three thousand to the east, and found their kind nesting in appropriate areas. There were condors then nesting in British Columbia and in Florida. Much has happened since.

The single downy chick in the cave high up in the canyon wall matured under the solicitous care of its parents, separated from them, found a mate, and became a parent himself. There were occasional mortalities among the chicks he fathered and some breeding years were barren, but enough of his kind followed to carry the species on down through the centuries. Subtle changes occurred, favored characteristics became accentuated while less vital ones faded in the genetic pool until the condor that was vanished and a new species evolved. There are hints in bones, to be sure, but the whole story is far too subtle for us to trace. We only know that it happened for we have the results. We understand it imperfectly.

While the condor altered itself and found new ways to survive, great changes occurred on and in the land over which it soared. The camels vanished, the ground sloth, the sabertooth, the mammoth and the mastodon followed. Bison became smaller, and probably more numerous, and the fabulous teratornis disappeared from the skies. A species of condor, two vultures, a caracara, an ancient golden eagle and four other eagle species as well faded away. One by one they melted back into time: the dire wolf, the short-faced bear, a tapir and a lionlike cat that had contended with the sabertooth, they all went. Somehow, though, the condor held on, for there was something there Nature wanted to preserve. Few new species appeared, for there was enough in the old collection worth preserving; but many had to go.

During the period of the condor's meticulously slow transformation, one new species did appear along the coast that would have profound meanings for the condor in the ages that lay ahead. Slowly, from the north, they appeared, not native to the land but migrants who had found good enough reasons to risk the coming. They could not fly, they could run only clumsily, and they had neither fang nor claw. They did have wit, though, more than any other animal, and that outweighed all the other advantages possessed by native species. Man had come bringing his greed, his natural bent for waste, and his open, often untempered hostility.

The first men apparently came across from Asia during the last ice age, when a bridge existed between the two continents. Strangely, they found the coastal regions of Alaska temperate and unglaciated. They moved down into British Columbia and from there south and east. We do not know for certain what they looked like, how they spoke, what they carried with them, or even very much about what they met along the way, but come they did to conquer first one continent and then another to the south.

The first relationship between man and the condor was mystical. The condor soared beyond the spear's throw, beyond the cast of the sling, so he was seldom hunted. He assumed, instead, an aura of legendry that became intensified down through the years. As the cultures of man throughout his range became richer, so did the tales and attributes that had the condor at the center.

The Tlingit people said the condor caused the thunder by flapping its wings, even by moving a single quill. The lightning, they claimed, came from the bird's red eyes. An angry condor, they were sure, was likely to create thunder and lightning until it was able at last to capture a whale to carry off to a mountaintop home.

To the people of the Tsimshian nation the condor was one who carried maidens away and captured the wives of other birds. In one instance, a single condor created a gale with his wings, destroying all other birds around just so he could have the wife of a woodpecker who happened, in this case, to be a thrush.

The Kwakiutl, the Comox, and the Nootka peoples as well feared the thunderbird in their legends as the great abductor. The Kathlamet called him *Ikenuwakcoma* and attributed to him similar tendencies. In the south the Hopi named one of the three branches of the eagle clan for the condor and favored the deity *Kwatako* with some of his characteristics.

The legends grew richer as the centuries passed and the condor, or thunderbird, became a favored subject of artists and craftsmen. In spruce-root basketry, on carved wooden poles, on cave walls, and in the design of pottery, the silhouette of the condor flourished. As the most masterful soaring bird in the sky, as an untouchable, unreachable image in black against the blue and the white, he became a demigod and as such would probably have survived across his vast range. But, other forces were due, and these were to be less benevolent.

From a different direction they came, across a different ocean. They had a greater greed, even, than the earlier men, a far greater sophistication, and guns. What the spear could not reach the musket could. What the dark-skinned man all but worshiped, the paler man destroyed.

It was in the year 1602 that the white man first acknowledged the condor in writing. A Carmelite friar, Father Ascension, recorded a flock feeding on a dead whale in Monterey Bay. His observations were included in an encyclopedic work published in Spain thirteen years later. It was a peaceful observation unlike those that were to follow.

1805: one killed at the mouth of the Columbia River by Captain Clark; 1806: two killed by a hunter in the same area. 1827: two killed by botanist David Douglas near the site of present-day Portland. 1845: an artist-ornithologist collected another. For each one killed and recorded, thousands were destroyed quietly, privately. Their huge eggs, often four and a half inches long and nearly three inches in diameter, attracted the collectors, and nests were robbed without regard for the fact that one condor lays one egg every two years. The collectors neither knew nor cared that Nature, in designing the condor's ways, anticipated a high rate of survival. Why else would a condor lay but that single egg in twenty-four months, while in the same period a tortoise lays at least two hundred, and a codfish twenty million? But the gatherers could not be bothered with such things. The eggs brought too high a price from museums and private collectors in the East.

The condor, of course, wilted before the attack. It had come too far, was too settled in its ways. By the dawn of the twentieth century, few existed north of central California, none east of the Rocky Mountains. But man's work was not done, even with that. Since then the vise has continued to tighten and today, of the thousands that once knew the skies from Florida to Vancouver Island, fewer than fifty remain, all nesting, when they nest, in a total area of less than 55,000 acres.

In Santa Barbara County, California, there are 1,200 acres in Los Padres National Forest, within the San Rafael Primitive Area. This is the Sisquoc Condor Sanctuary. North of Fillmore, in Ventura County, there are 53,000 acres designated as the Sespe Wildlife Area, and here are most of the known condor nesting sites. This is what man has left of North America to the condor. This is his share. The bird that flew in ancestral form before man even knew land lay in this hemisphere, the bird that fed its young on the flesh of imperial mammoths and sabertooth cats, this is what has been left to him. It is not at all clear that we will allow him to keep even that. Covetous eyes look in his direction even now. Our greed is not sated, the condor's last chapter may be about to be written.

The promise made to the future by the *Archaeopteryx* one hundred and fifty million years ago lives on in the condor today. It remains to be seen if man will allow the promise to be kept.

❧ 12
Part of a Grander Plan

From
Source of the Thunder

Each stage in an animal's life story carries it through elements of its habitat, and even in relatively restricted areas, such as that still utilized by the very few surviving wild California condors, there are changes. An animal's life, of course, is one adventure after another and for the reader who has opted to peek inside that life the habitat should be an adventure as well. An animal is never free of its habitat and its twists and turns and secrets, and the reader of that animal's life story can't be free of it either.

Again, it isn't necessary to publish seed catalogs of what an animal sees and eats and uses in other ways (as nesting material or roosting opportunities, for example). But it is the writer's task to slip this in a little at a time so that at the end of the tale the reader will not only know the animal but the animal's place in the grand scheme. If the author and the reader together have not seen the animal and its place as part of a grand scheme they have not understood the story at all.

The cycle of seasons and years spun on, carrying everything in the condor's world with it. His journey into maturity was as inexorable as the planet's course, and each was a part of the other. His link with the world and the cosmos was as firm and real as his ties with the past. Only his lines into the future were uncertain, for his species remained poised on the brink of oblivion throughout the years of his immaturity. Men of exceptional goodwill and with an extraordinary capacity for moral responsibility voiced their concern, and his plight was lumped together in a limited public conscience with the perilous position of a thousand other species of wildlife around the world.

Although the condor as an individual bird was known to anxious men who watched him through powerful field glasses, recording his travels and noting his growth with satisfaction, and although his range lands were patrolled by salaried men who had vowed to protect his kind from willful intrusion, he knew nothing of it. He was locked up within the limited daylight of his bird-size brain, responding to each moment and each experience as his species always had. For him there was no census, no recorded rate of reproduction or attrition, only food and its getting, wind and its value, storms and their hazards. On the undersides of his wings there was a hint of white, a promise of large triangular markings that would soon appear and announce his capacity for reproduction. When that time arrived, everything that had gone before would have a purpose.

Few days in his life were uneventful. The seeking and claiming of food had sharpened his powers of observation, increased his skill at low-level flying, and given him a growing air of authority suited to a male nearing his maturity. He was no longer chased off carcasses but held his own against all but a few patriarchs in the condor world. Like all of his kind he continued to show deference to some other species, as he always would. The golden eagle, the great *Aquila*, was his master on land and in the sky. The coyote, the mountain lion and the bobcat continued to intimidate him and his reluctance to challenge them when they appeared was a factor in his survival.

Cattle, deer and horses remained high on his food preference list. Man's increased skill in animal husbandry and veterinary science made the supply of carcasses from domestic stock somewhat smaller than the condor's antecedents had known but still there was enough. Small carnivores trapped for their skins, shot for sport, or destroyed as vermin provided occasional food but their naked carcasses were more often discarded in places unsuited to his feeding habits. Although he occasionally fed on a coyote or a bobcat he could perceive no irony. Such intellectual sophistications were beyond his reach. He accepted food as food and didn't concern himself with its source or the reasons behind its availability to him as carrion.

Technically a bird of prey, although he did not actually prey on living animals, he worked under a decided handicap. Whereas the owls, the hawks and the eagles and their kin could carry food off to high and secluded places to eat in safety and at their leisure, he was forced to eat where he found his food. The other predatory birds have talons and powerful feet that can close over objects of reasonable size and carry them off. He was flat-footed, with feet designed by Evolution for walking and standing only. The legends of condors flying off with everything from men to whales in the talons they do not have are legion and many believe them even today. In simple fact, a condor cannot fly off with a mouse, for he cannot even grasp it. When feeding he can hold a carcass down by pressing on it with a foot while tearing at it with his beak but there the utility of his feet stops. This simple fact requires that he work harder, much harder, than other flesh-eating birds. There were days when the young condor went hungry.

In his third year, while flying low through a sparsely wooded valley, he spotted the carcass of an ancient horse near some brush on an easy slope. He curved in a tight circle and flew over the carcass a second time. There was a dead tree a few hundred feet east of the find and he settled onto its top branch to survey the scene. There was no movement in the valley except for a few lizards in a rock-slide but there was a vehicle, a jeep, parked several hundred yards down the slope. He had often seen these small dusty machines and had seen the men who controlled them, but had never learned to associate them with danger. The vehicle was immobile, tight against a large bushy outcropping beside a small stream. He worried about it for a few minutes, preened a few feathers to dissipate the energy his anxiety occasioned, and finally decided it was safe. He kicked free of his perch and flew straight to the carcass.

He landed easily and walked to the horse. He spread his wings, hopped once, and landed on the animal's bloated belly. He couldn't note with any understanding that a bullet hole was centered on the white flash on the grizzled old horse's forehead, nor could he observe with comprehension that the animal's hide had been slit in a number of places to ease his task. He missed, as well, the tire and drag marks that showed plainly that the dead horse had been dragged up the slope behind the jeep. The most important fact he missed was that a crude but effective fence had been set up around the carcass, connecting the obstructive brush and rock outcroppings that littered the area. He had landed in an arena, designed and built for a special purpose. As soon as he had eaten his fill he would automatically be the prisoner of the men in the jeep at the bottom of the slope.

Horse flesh was a favorite food and the slits in the hide of the newly dead animal made his feeding easy. He glutted himself, working rapidly against the intrusion experience had taught him would not be long in coming. In less than fifteen minutes two other condors approached the area and, made secure by the sight of the younger bird eating, circled in on the carcass without first landing on a vantage point to survey the scene. The carcass was large enough so that there was no appreciable squabbling. The three birds fed together and became so engrossed that they didn't notice the first sign of movement down by the parked jeep. Two men started up the slope, one of them carrying a canvas bag.

When the men were about three hundred feet off, one of the older birds, the second to land on the carcass, saw them and grunted his alarm. He turned and ran awkwardly down the slope to get the wind under him. Before he could lift off he ran into the fence and collapsed against it. He sat there, his wings askew, looking foolish and ineffective. The other mature bird and the young condor followed his example and piled into the fence near him. The movement of the men up the slope panicked the birds and in urgent haste they righted themselves and hobbled up the slope and then started down again. Three times they piled into the fence, unable because of the weight of the food they had consumed to lift off in the limited space allowed them. They were trapped by their meal.

The sight of the men climbing over the obstruction into the penned area further terrorized the three birds. They careened into the fence, tried to crash through bushes too thick to allow the passage of a sparrow, and collided with each other. The young condor was no more at a disadvantage than the others. Even their many years had not taught them how to react to this situation.

What happened next was a blur for the young bird. He saw two of the men closing in on him, blocking him into a corner of the pen. As he rushed to get past them, a cloth bag settled over his head. In the dark he was helpless and came to a stop. He felt firm hands grasp his wings and pin them together over his back, heard voices, felt himself lifted off his feet and placed on the ground upside down. He grunted and moaned but quickly stopped his struggling. There was no direct pain associated with the men's activities but the fear he felt hurt almost as much. Even if his head hadn't been covered he would not have understood the long tweezers that prodded in between his feathers in search of parasites or the thermometer that was slipped into his vent. The few feathers that were pulled would not be missed, and the injury to his dignity would not be crippling.

His misadventure was soon over. He was left lying in the dust and when the bag was removed from his head a few minutes later the men were starting away down the slope. The fence on the downhill edge of the pen had vanished and the other birds stood nearby, blinking foolishly in the sun.

None of the birds could appreciate that the examination they had been subjected to could supply science with information that, in combination with other painfully extracted data, might work toward the survival of their species. For them it had been a thoroughly distasteful episode. All wild animals, and a good many domestic ones, are badly frightened by any hint of restraint. Complete freedom of movement is the very essence of being wild. There is an atavistic dread in being held immobile. Too often in a world of carnivore-prey relationships it means being caught, probably to be eaten.

It didn't take the birds long to recover their composure. They were soon off down the slope, grunting with each wing-flapping hop, and then were off, comfortably resting on the rising cushion of air that bunched up beneath them as they rose over the ridge beyond the stream. They didn't look back.

Having shared the most disquieting adventure of his life with the two adult birds, the young condor stayed with them. They rose high over the next valley, strung out in an approximate line, and climbed over a mile into the sky. Twenty miles east they settled into a high meadow where a clear stream ran and bathed there and drank quantities of the near-freezing water. After their preening (the two adult birds nibbled each other's feathers but didn't invite the younger bird to participate) they rose off again and flew north. A storm front was pivoting in from the west, starting a southward swing. The birds rose to seven thousand feet to avoid some particularly disturbing turbulence and as the young condor beat forward his temporary companions veered west.

Then, seven thousand feet above the mountainous earth, with black clouds building up on all sides, out of sight, scent and sound of all of the billions of living creatures with which he shared his planet, the young condor was once again alone. There is no solitude on earth the equal of that above a mass of storm clouds. There silence is a sound, nothingness an object, and aloneness a social state that crowds in on a single organism and possesses it. The young condor began his descent, down through the clouds since there were no passages open in view. No one saw him come down, no one saw him settle in the dead oak in the valley and huddle against the storm. Millions of other creatures felt the wind and the rain and heard the symphony of their fury, but no one and nothing thought about the condor alone in the tree, a bird on his way to his maturity, carrying within him the treasure of the eons, one fiftieth of the genetic potential of his kind left on earth.

❦ 13

The End Is Foreordained

From
Source of the Thunder

Death is a part of life. A dead animal is as intimately a part of its place as a live one is for it is then the creature repays its chemical debt. All of the chemistry that an animal represents at the time of its birth and during the years of its feeding are borrowed. Through biological processes inherent in breathing in and exhaling, eating and excreting, an animal remains part of the chemistry of its place from birth to death, but it isn't until death that the large outstanding balance of that bill is paid off.

If looking at death as something other than biological and chemical commitments is sentimental, then most readers and writers are sentimentalists. There is nothing wrong with that. I don't know how many people have wept for a condor but there is no reason not to, not as long as we weep for our horse, our dog, and our cat. It is all much the same kind of thing. Being able to weep for an animal, after all, is a hallmark distinction of our species. It is as distinctive as the shade of a robin's breast, the wonder of a condor's wingspan, and the great cat's spectacular ability to roar.

At twenty, the condor who had soared nearly a half a million miles over his mountainous range was tired. He had lived seven thousand and three hundred days, he had pressed himself to the bodies of two females, one of them his own offspring, and had produced a desperately needed crop of successful chicks. He was immortal. He had known peace, and he had known fear. He had cheated death more times than we can know and had thereby protected his potential, keeping it from harm until it was used up. His had been a successful life and he had lived it in the sun.

It was the breeding season again but the reproductive equipment within his body failed to swell. When he turned away from his daughter-mate and refused to display, she turned away from him and flew off across the canyon to where a recently matured bird was waiting. She could never think of the old bird, for that was clearly beyond her, and she would not recognize or acknowledge him when they met at bathing sites or on carcasses they found in common.

The old male spent more time on his perch now than he did aloft. There were days when he didn't bother to seek food but sat hunched down, his eyes half-closed and his beak slightly agape. He looked surprised, bewildered, as if he couldn't understand why he was being kept waiting.

On some days, when the sun was warm and the breeze exactly right, he seemed to revive, to soar again and to exert his position at the feeding ground. But those days became fewer and fewer. He was becoming forgetful and there were days when he didn't bathe. He became untidy and his feathers fell, leaving bald patches that were not only unsightly but interfered with the effectiveness of his complicated flight mechanism.

His decline was steady, a constant thing like everything

else in his life had been. His cells were losing their memory, forgetting the things they knew when he and they were young. His brittle bones ached under the strain of movement and he limited himself as much as he could without actually starving. Younger birds were now pushing him off carcasses and he seldom ate his normal allotment of two pounds a day. He soared unsteadily over longer distances than were necessary, just to find a water hole that was not being used. Although he no longer bathed he did require water to drink and he preferred places where he would not be hissed away and insulted.

When he landed on a perch it was an unsteady maneuver and he no longer was willing to fly unless the wind was exactly right. He was ragged and sad and he had lived a little too long. And then, one morning, he was missing from his perch. No one noticed or cared, but where a condor had shrugged painfully into a sleeping posture at dusk there was a bare branch at dawn. The brush below the tree was thick and tangled and his body didn't show. But a coyote had her den site there and her cubs growled with mock anger as they tugged the carcass apart in their learning games. A single feather from his breast was left on exposed ground and as a soft eddy of wind began to move it around, a horned lark dropped down to collect it for her nest. In the nest she laid three olive-buff eggs sprinkled with drab and lavender, each of which would produce a bird far, far more beautiful than the condor had ever been.

· 14

The Underwater World

From
Sockeye

No land setting is as foreign to the average reader as the world under water. A puddle or an ocean, it is much the same thing. All perspectives change, senses are smothered by the heaviness of water compared to air. The creatures there respond to different signals (although distinctly not different imperatives), as what we think of as distorted vision and sound and near weightlessness replace what we experience up here in the world of air. It is a special challenge to write about life under water and to try to make a creature that lives there sympathetic. It is even more of a challenge than trying to make a great, black vulture a hero, as I have tried to do with the condor.

But under water is a place—the place, in fact, where we came from ourselves—and the animals that live there now are animals as surely as we are and the elephant is and the cockroach. An animal under water may be foreign to us, but it is nonetheless an animal in its place and there can be no question of understanding anything about that animal until the place is first a reality for the reader. It is an interesting undertaking for a creature like an

*author who has spent only seconds under water, no more than that,
to try to recreate the life of an animal whose entire life is spent
there.*

The lake itself is above Cook Inlet north and east of the
Alaskan Peninsula. It is north, too, of the fabled Kodiak
Archipelago. Its name is Coppertree, and the river that
empties into it above and flows away to the sea below is
called Coppertree Creek. No one knows where the name
comes from. The lake is nestled in an area of alder and
birch, of evergreens that finger the sky and then suddenly
relinquish the land to areas of scrub. There are blueberry
patches, and they plus the guaranteed salmon run every
year make it a paradise for bears. There are brown bear in
the vicinity for most of the year, gruff shuffling giants, rude
and coarse; and there are moose and fox, of course, fox
hungering after the leavings of the sloppy bears. There are
black bear, cinnamon and black, actually, but generically
black, and there are wolf and deer. The smaller animals,
weasels to voles, number in the tens of thousands around
the Coppertree's perimeter and along the banks of its river.
Osprey fish; the northern race of bald eagle fishes, too, and
also plays the scavenger taking whatever is to be had. Ravens, crows, nutcrackers, jays and magpies vie to outdo
each other in rudeness; lesser birds come and go with the
season. The sunlit hours ride a wide-swinging pendulum
and the lesser birds, the smaller perchers, are dictated to
by seasons' march and respond in order to survive. Coppertree is south of the Arctic yet north of the temperate
surprise of southeastern Alaska. It is an almost-world, a
place of fog and rain and then sudden balmy calm. Harsh

and cruel, gentle and caressing, the world of the Coppertree does what it wishes and its schedule of events is its own. As sodden as the forest and the sky might ever be as they hang above the Coppertree, the salmon fry are restricted forever to the world beneath the surface of the lake and the river systems that support it or are sustained by it. Occasionally, in the final frenzy of their spawning run, the hens and cocks might thrust across a spit of land, a forgotten intrusion of sand and gravel. But for almost every moment of their life they live in the more temperate and temporizing world submerged.

In order to survive this next great phase in his life, *Nerka* had to enter Coppertree Lake at a time that coincided with several phenomena. There would be enough tricks of chance in his time as it was without allowing basic rhythms to run high risks as well. Those had to pulse with the planet, with the weather enveloping all and with the beat and march of other forms of life.

The first coincidence was that *Nerka* as a fry had to enter the lake at a time when the ice was breaking up for it was then that optimum conditions of temperature, oxygen, and prey were available. The top yard of water had been brick-hard for months, but as May approached it grew thinner by the day. Water from the uplake Coppertree Creek flowed across the warming surface and pockets and dents began to form and then collapse inward. Sheets of ice deteriorated into spines and spicules and these, too, became frayed by the eroding warmth of earth-tilt and time-change. Each day was longer than the one before and each promised more sun to follow. The color patterns that flowed across the fracturing world of ice were at places in the spectrum for which words have not yet been invented.

Rain, when it fell, came through a warmer sky and chewed at the lake's icelock and fragmented it until hunks broke loose and began to find the downlake river flow sloughing away at the edges as they went. Winds came and went, ice blocks rocked and pushed and their rudeness further eroded the hold of a winter gone and dead. And then the lake was free. The Alaskan months of chill and harshness had come and gone again, bringing life and death in equal shares. The animals that now came down to the lake to drink were the best of their kind and they carried the genes of the future. The birds that now worried over nesting site with song and posture were also the better of their species and they, too, were bridges from the past to the future, the bearers of genes and promise and evolving wonder.

The second coincidence of *Nerka*'s coming to the Coppertree Lake was the proliferation of life there. Day by day the plankton population was increasing for it was spring beneath the surface as well as above. Small crustaceans, small enough for the fry to eat, swam in profusion and were there to build the ounces of salmon flesh. The insects that spend their infant stages on the lake bottom and in its middle waters were there, too, by the billion and *Nerka*, as he grew, could rise to the surface where aerial insects would sometimes fall and pock the surface. They could be pulled below and used as well. The planktonic crustacea, though, copepods and the cladocerans like Daphnia and Bosminia, constituted *Nerka*'s staple diet.

It was a time when the level of the lake rose and spilled out across its own grassy banks. Ice and snow further up the system rushed down the creek, swelling it and carrying the fry to the waiting lake below even as they swelled the lake itself. For a time, during those first days, *Nerka* moved among the blades of grass that would soon, in the summer months, be land again. Where deer would browse *Nerka* now hunted. Where grouse would wander and strut *Nerka* prowled, an infant seeking to eat other infants. Other fish prowled there, too, and only infinitely unfathomable luck of numbers allowed this fry to survive. He was one of millions that would live, of course, around the entire northern Pacific periphery, but billions of others died. Trout and carp and other creatures that share the lakes and rivers with the sockeye fry seek to build their flesh too, for it is their spring as well. *Nerka* darted for prey and darted equally often for shelter. The future land grass was useful, it provided cover almost as thick as the redd in which the egg had become an alevin even before it had become a fry.

Once the fry have taken up their truly pelagic existence they do not distribute themselves equally throughout the world of the lake. Although in the winter months they will sink away from the ice and seek deeper mild waters and basin hollows, in their first spring and summer they remain close to the surface. They live as part of an enormously complex limnological community and circulate with it. Masses of planktonic crustacea circulate slowly with currents within the lake's boundaries and the young fry stay with the moving, undulating feast. In the darker hours of day, early morning, late afternoon and days of hard rain and lowering skies, the fry rise to within a foot or two of the surface and at times come all the way to the surface itself. When the sun is highest and the summer most nearly keeping pace with promise, the fry circulate with their banquet from fifteen to twenty feet down. The edges of these masses of life are frayed, of course, individuals and small clusters drift up and down and away, but this is the general pattern. It is less casual than it may seem for the needs of all species are met.

Instinctively each sockeye fry in the lake concentrates on eating. The larger it grows, the more quickly it achieves fingerling size and the larger the fingerling it becomes, the better its chances for survival. In the early pelagic days of May, the fry each consume three-tenths of a milligram of food a day. By midsummer that rises to over thirty-one milligrams each twenty-four hours, and although the winter sees the figure reduced sharply as food supplies dwindle, April of the following year will see the individual needs reach toward forty milligrams between sunrise and sunrise. The population of the Cyclops and Daphnia, the Epischura and Bosminia in the lake explode to meet the challenge. Like the cosmos with its uncountable billions of units of light, the minute crustacea expand geometrically and *Nerka* was to feed along with the rest. He survived his weeks in the redd as egg and alevin, he challenged the uplake river reach as a fry and then rode the current to the lake with the ice breakup in springtime thaw. He skirted danger on gravel bed and among weeds and reeds never known to him before. Sunken objects held safety and peril in equal share. He learned to dart and duck and feed while avoiding hazards that grew by the minute.

Nerka now was part of a perfect circle and at any moment he could be plucked away to dissolve as the food of something else. His mission was clear. He was entered in the race to see if he could survive long enough to pass across a redd near the one where he was hatched. He was to live long enough to spread his seed across the gravel and the unfertilized eggs that had begun to settle there. Then his fate didn't matter. Chemistry is a leisurely element in life to which all must come back in the end, so never is there any kind of rush. It is what comes before the chemical return that matters and that was the cycle into which fate had decreed *Nerka* must enter. A future male salmon, he was a spark of life in a spark-filled world. Other sparks filled eagles and bears, others birds and men. Whales, too, are sparks and all that are smaller than that. The spark is the life and the purpose, the chemistry the framework that holds it all and in the end collects it all, all but the spark and that—the destiny of the spark—is the greatest mystery there is.

• • •

Even as *Nerka* pushed forward through each day toward his destiny, his private history was being written thousands of times, repeated over and over on each scale of his body. If eyes of man were ever to examine this single fish, the days of his life could be read.

Throughout his years circuli or sclerites were laid on his scales until they gave the appearance of ridges. Under a microscope they would be as clear as the growth rings of a tree. In summer, when his growth would be rapid because of the abundance of food, the circuli would be far apart, widely separated as his growth increased. In winter, when his rate of growth would slow, the markings would be closer together. They are known to men as winter bands and their count is the recorded age of the fish. When he finally moved down through the last rung of the sweet water system where he was hatched, *Nerka* would then be plunging ahead toward maturity. The circuli would be less delicate and their grosser nature would be easy to see. A knowing eye could tell the seasons in the lake from the seasons in the sea. If he was to be an average fish, his years would number between two and seven. The time allowed for the full and miraculous life of his spark was no more than that—and no more was needed. All that a salmon is or can be is locked in step with the lives and styles of the other creatures with whom *Nerka* would have contact. The meshing meant that all would survive, all species at least, and despite appearances there was no capriciousness. It was locked, fixed, productive and whole—chemistry taken from an inanimate system and made alive for a while. In the end all would meet an inevitable common fate with the stuff of their life returning for use again. We know of no perfection beyond this, and nothing more miraculous.

As the young salmon moved out into the lake, as he became a viable element in the world of the Coppertree, his survival would depend in large part upon his ability to react; and in order to react he had to receive signals. Although his brain was small, too small and too simple to allow for much learning, it was superbly attuned to the senses which in turn were exquisitely receptive to all that went on around him. His eyes, for example, were remarkably like those of man although water is a far different medium, a far poorer conductor of light than air.

Nerka's eye was a camera. Light rays entered through the lens at the transparent center of the eyeball. From there they were directed with precision onto a light-sensitive screen, the retina. Where the human eye has an iris or diaphragm in front of the lens to control the light coming through, *Nerka*'s eye had none. The iris was fixed. But there was no loss of control. A perfect system had otherwise been evolved. In the screen or retina at the rear of his eye there were two kinds of receptor cells—rods and cones—and each sent their messages to his brain. The cone cells could receive color and were at least thirty times as sensitive as the rods, black-and-white receptors only. During the day, when the sun played across the lake's surface, the cone cells were in use; but as the light level fell, as evening crept in across the surrounding hills and fingers of dark slipped in between the trees, then a miraculous thing would occur. The cone cells that had been active since dawn would begin an actual retreat, back into the deeper and darker layers of the retina, while the rod-shaped cells moved forward. During the bright hours of the day the rods remained hidden in that same deeper and darker tissue, protected against the light that could injure them.

Nerka was nearsighted, for nowhere would he ever be where visibility would come close to equaling the world above the surface. In the most crystalline water 99 percent of the light is lost at twenty-seven feet. When storm toss and high river flow create regions of murk, little light can penetrate beyond a dozen feet. The plankton fields upon which *Nerka* grazed further reduced the light available to him and his visibility range ran from five to a true optimum of forty feet. In fact the food that *Nerka* hunted, the minute crustaceans and insect larvae, acted on him as fog and smoke does on the vision of man. They blocked and scattered the light coming toward him in the water. If his world had been perceived visually alone, it would have been confusing and even more rich in hazard. As it was, his visual judgment of his world was based on two phenomena, movement and contrast. A flash, a movement, a change of light values, a shadow, a silvery belly, these made the visual quality of *Nerka*'s world subject to interpretation. Without

them it would have evened out into a foggy sameness beyond his discerning, beyond his reading.

There was another strange aspect to *Nerka*'s visual
world. It was the sameness of it, the constancy of cold light.
Fish do see color, of that there can be no doubt, but water
is cruel to red. The red-yellow-orange range, the warm
end of the spectrum, disappears almost as soon as a light
ray penetrates the surface, and what do manage to sink
deepest are green and blue.

On calm days when *Nerka*'s body was oriented upward,
the underside of the Coppertree's surface was a silver mirror, and when the water was shallow even a reflection of the
bottom might play there. There was this silver and the
silver signals of other moving fish, but even silver is cold.
That was the world he saw and the only world he could ever
see—fractured light, bouncing light, refracted streams
and jets of silvery blue and greenish-silver in a world of
constant motion. It was a cold world of cold light and it was
for this that both *Nerka* and his eyes had been fashioned.

Nerka was far more dependent on his sense of smell than
man can ever understand. Unlike his counterparts on the
land his nostrils did not connect with his throat. Instead,
each ended in a chamber lined with sensing cells so utterly
fine we seek in vain for a valid comparison. Over half a
million receptors per square inch interpreted the minute
chemical messages that were pulsed inward by fine cilia.
They moved in rhythm to assure the constant flow of information into these chambers and from there to the brain.
Indeed, man has always wondered over the indisputable
fact that the largest part of the fish's brain is given over to
this single task, the interpretation of and reaction to the
sense of smell. *Nerka* could detect traces of gas and more

solid matter far beyond our detection. Like a finely tuned instrument, he locked his visual world and his chemical world together and used them in tandem to hunt and to successfully elude hunters.

From the moment he emerged from his pink egg as a pink alevin—perhaps before that, for all we know—*Nerka* had another use for his sense of smell as well. Perhaps his use of this sense is greater than that of any other animal on earth, or at least so some people believe. He began at once developing a chemical memory. He began a process of imprinting so beyond comprehension that we sometimes long to deny it on that basis alone, its strangeness to us. When *Nerka* left the Coppertree and migrated through its lower reaches to the sea, he would program his memory to unwind again with incredible precision. No matter how many thousands of miles he might wander during his sea-years of growth, this memory would one day at least in part help him to smell his way home again and do that well enough to rest at last over the gravel bed where he had been spawned and hatched. For *Nerka*, resident of what to us is the utter sameness of the water world, no watery place, no liquid track smelled the same as any other. And his brain, that brain we call small, could remember the infinite details that built that fact with shreds of scent. Try as we will, we can never really understand how that is done. If we could record with the same precision and permanence the product of our sensory intrusions as what *Nerka* learned from his streams and his lake and what he would later learn in the avenues and alleys of the sea, what manner of intellect would we be? But we think, we conjure, we calculate and we learn, and so we cannot spare the proportion of brain *Nerka* could devote to this aspect of his salmon existence.

There was more. *Nerka*'s sense of hearing was no less stunning. His ears were buried deep inside his head and were not exposed to the water at all. There were no flaps, no holes and no eardrums as we understand them. Sounds conveyed to him in a turbid and even polluted world could be interpreted as they passed through his skin, his flesh and his bone to reach the ears safely placed away from trauma. Miraculously he was able to distinguish the good from the bad. All his life, as if by magic, he would be able to tell the sounds of those that hunted him from those of the creatures which he had to hunt in order to live. And all of this came with him in the egg. He learned practically none of it himself. It had been learned for him.

And yet even this was not enough. *Nerka*'s hearing was enhanced to an incredible degree by tiny pods strung out along the length of his body. These cupolas were lined with sensory hairs and connected by tubes to the surface of his skin in two lines running the length of his body from his head to his tail. Called lateral lines, they enabled him to detect even the lowest frequencies. The tiny canals picked up the lowest thudding disturbance, and when the information of that was coupled with the actual sense of hearing, *Nerka*'s ability to pinpoint source was beyond anything we can ever even approximate. Up to thirty feet away *Nerka*, stationary and attentive, could read the sounds of his world. Beyond that his lateral lines would fail him; and dependent on inner, hidden ear alone, he would have to swim a search pattern.

As *Nerka* reacted to the world around him, he often did so in a rippling sensory pattern. Either sound or smell might first attract him to prey or danger, and one or the other might at first predominate in worth. The one, though, would soon overlap and play off the other until sight became of value. Then, with sight, smell and sound in harmony the fish could become a truly responsive mechanism. Another chemical-reception system, taste, also helped him monitor his world. Unlike man, his taste buds were not limited to the interior of his mouth. They occurred as well on his lips and snout. He could taste something by simply brushing it and need never endanger himself by ingesting something he could detect as bad.

Man has long wondered how well fish can feel, respond to texture and immediate pressure. This is not yet measurable by means we know, but we assume that it is reasonably acute. It worked well enough for *Nerka*, and by this means, too, he judged his world and used it more quickly than he would allow himself to be used by it. One touch we know was sure and swift in value. *Nerka* could judge temperature and even navigate by it. He could follow beacons of heat and cold, currents that led to and from where he chose to be or not to be. Ribbons of temperature ran through his world, a warp against the woof of silvery green and blue, against the texture of sound and pressure sensed along the length of his body and against the fabric of smell and taste. He used it all without knowing and without the ability to care. It was there in the egg, it was there while he busied himself with his yolk sac content, and it was there as he grew. It would always be with him. He was a spark and these senses were the contents of that flaring, flashing miracle of life. All of it would die with him. All would dull and fade as his life faded; but that was later, after the lake, after the sea.

.　　.　　.

Nerka's life was one of movement. He swam to get from one place to another; when in currents and eddies, he swam to stay in place. He moved both toward food and away from peril. He was designed, totally, every part of him, to move through a medium many times the density of air. Unlike the surface animals of the land, very much like the birds in this one regard, he lived a life that was as vertically oriented as it was horizontal. He moved forward no more often and no less than he did up and down.

As a mature cock *Nerka* would be able to explode forward at fourteen miles an hour. He would not be able to sustain such a speed, but he would be capable of achieving it. As a fry he could come nowhere close to that, but he was still, in miniature, the swimming animal he would always be. His coordination was as smooth as the passage of a season and as true to set form. In a forward movement, either in pursuit of a minute animal or as an escape tactic, he would go from a stretched-out streamlined form into a virtual semicircle in less than one-thirtieth of a second. Stroking hard he could arc his body until the same maximum curvature was achieved in the opposite direction and that in less than one-twentieth of a second. His velocity now close to maximum, he could repeat the stroke again in one-thirtieth of a second and then glide with his body extended. This could be repeated as often as needed. It was too swift properly to be seen, but the results were clear.

As he moved through the water, he created his own complex pattern of eddies. Upstream falls and rapids were no more dynamic than these, and no more complicated. The fact that they were in miniature simply suited the scale of the animal itself. As he grew, the currents and eddies he would create with each move would increase and blend in with the billions of others created by all the other creatures in the lake or the sea. No one has ever reckoned the role of these in the movements of the cosmos, how such energy blends with all others so released and how they play upon each other. This much is certain, though: with his exquisite sensory equipment *Nerka* could read those currents created by the creatures around him and judge their worth, judge their peril. Even as he created currents and sent water particles swirling against each other and in concert against other forms of life, he interpreted second by second those that returned to him from other sources. As he moved with other fish, they communicated by the gentle pressures that the movement of each exerted on all others. When we think of the fish, we must think of this, the degree to which it is locked in step with the cosmos as well as with smaller wonders. It is all one, and *Nerka* a part of it.

By the time *Nerka* was ready to go to sea, 96 percent of the young produced in the spawning run of his parents would be dead. Many of them would die in the lake, in that two-season residence in sweet water above the sea. A principal and deadly foe of all salmon fry is the Dolly Varden char. A fast and greedy fish, it seeks the sockeye fry and at any one time may have as many as ninety of the infants in its stomach, ninety of that year's crop.

Nerka had fed well on a passing raft of crustaceans, a planktonic feast circling the edge of a lake basin with an easy current. The current was born in the creek above the lake, born where *Nerka* had been spawned, and it blended with others coming in above and below the lake. It drifted around in lazy circle and carried flotsam with it. Occasionally small treasures would be snatched from it by countering currents and intersecting eddies, but most of what it claimed traveled with it, round and round. *Nerka* had joined it as he often did and drifted in among the motes that sparkled and caught and kicked free cool silver-green light. Bubbles drifted among the motes and these were gems, too, gems that hung and glittered mid-water. *Nerka* browsed on the eternal feast and tasted and felt and sensed as much as he could of his ambient world. He was growing and that was the present mission. Now he competed for food with others, not always his own kind.

As he dropped back from the drifting cloud of crustacean prey he slowly let himself down deeper, behind a sodden tree, to where some weeds grew in small clusters. The water was no more than a tall man deep there and the sun splashed down above, bright and yellow against the Coppertree world. A slight breeze blew and chopped the surface and broke its mirror. Through the fractured shards of water the light fell and came glistening against the pebbles and the boulders toward which *Nerka* sank. It was an hour when the water was full of jewels and the bottom sparkled like a pirate's chest of booty newly opened to the sun.

Then, suddenly, there were signals. *Nerka* turned to find their source, to seek the meaning. His interpreting of them had to be almost instantaneous or his life could end. It generally was that swift.

A greedy Dolly Varden char had drifted in with a small bottom eddy and hung among the weeds only feet away. It had fed all day and killed three-score sockeye fry. Hunger was less the motivation than habit. It had been eating and would continue to eat until an unknown calling prompted it to drift away toward a deep place to digest the lives it had ended. Suddenly it, too, sensed the encounter. As fine a creature as *Nerka* was, he sent signals that others could discern. Even as he was sensed, *Nerka* sensed the Dolly Varden turn toward him amid the weeds and make the first tentative flick of its tail. The large and powerful body started forward in a first glide. In a second or less it would lash once and as designed would be in full pursuit. Mouth open, greedy jaws apart, it would strike and gulp at once and *Nerka* would vanish like all the rest. No chewing, no biting, just a single gulping motion, an opening and a closing, one cycle and all that *Nerka* was or promised to be would be over.

But *Nerka* had been well supplied. All the traits and powers destined to the sockeye breed were in him. At the first tail movement a wave of warning struck his side and slid along the microscopic tube endings from head to tail. The canals opened and the sensory cells read true. There was no need for *Nerka* to understand or linger on the signals; they bypassed such wasteful measures. *Nerka* moved, too. Just as the human hand releases a hot poker before the human brain can think of heat, so the fry reacted before any other process could occur. The Dolly Varden char passed on its profitless search and *Nerka* hovered low near where a boulder overhung a shiny place full of dancing chips of sun. It was the kind of darting existence the salmon would live for two years in the lake. At each stage of his life and growth there would be different fish offering the greater threat, but there was no day, no night when this evasive action would not be required scores of times. By rote, without powers of thought, without knowledge of fear in any form we can understand, but by merely being, he learned the safe places, and there he hung and lingered as often as possible. When necessary he ventured forth and drifted with mists of food, always dropping back toward shelter whenever sated. It was during his rise toward food and fall away from it that he was most vulnerable. Still, somehow, *Nerka* lived on. Each day his promise increased, for each day he grew and came closer to that day when he would be a cock salmon needed by the Coppertree above the lake where it could all begin again.

. . .

And so that first summer passed into fall. *Nerka* was progressing rapidly through his fry stage and could be properly called either a fingerling or a parr, the former perhaps more nearly correct since he was a fish of a Pacific species.

It had been a time of growth and of honing. He had become larger and faster although still minute and vulnerable as the community of the lake was viewed. All around him fish spawned at the same time and even by the same female and fertilized by the same male had died. Some had been infested with minute parasites and so slowed in their reflexes that they were the more easily taken. Others had fallen prey to fast trout and greedy Dolly Varden through no fault or weakness other than that to which all living creatures are subjected—chance and perhaps piscine fate. Still, enough, as always, had lived and among them was *Nerka*. He was a complete fish, full of those things that made his kind a species, different from all other kinds. He had passed the seemingly impossible test of a single spring and a single summer; now a fall and a winter lay ahead.

The avenues that run from the Coppertree system to the Arctic are broad and open. Cold pours down like a fluid substance. One day summer is gone and leaves are turning and falling. Branches of green become red, or yellow, and then they are brown and then they are bare. Small, tightly furled buds contain the assurance of another year, but for a season it is over.

The cold ache from the North that fell across the land

caused radical changes in the lake as well. Water can remain liquid for only a brief band of the heat spectrum—from 32 to 212 degrees Fahrenheit. Above that it is steam and below it is ice. And it was ice that began to exert its force upon the world of the fry, ice that had always been there as one of water's amazing potentials.

Water, as a liquid, consists of a double molecule, dihydrol. When it becomes ice it is triple in form and is called trihydrol. Even water contains some trihydrol, though, and as its temperature drops the percentage of those molecules increases. At first these triple molecules with all the threat they hold are in solution, but as the air above cools and sucks heat from the water body, the density of that solution increases. At freezing point the triple molecules precipitate and an almost-solid exists where once there was a liquid.

The first ice was not crystalline, but appeared as small disklike particles. They were, in fact, neither liquid nor truly a solid yet, rather a colloid, the mysterious in-between state of matter. The colloid particles grew rapidly as the air temperature dropped. Snow was falling and flakes were dying on the surface of the water. In hours they would be able to accumulate there. Below, the fish retreated.

In the water another intermediate phase came and passed. The ice was half-colloid and half-solid and then, almost suddenly, it was crystalline. The Coppertree had begun to freeze over. There was a buoyant mass, cloudy, less subject to agitation than water itself, and it floated more than half submerged. A sodden, as yet insecure mass shut the lake away from the sky.

Near the intake from the Coppertree stream above the lake the ice acted like soggy canvas, rising and falling and sighing with the current. Toward the center of the lake the weight of it quieted the effects of the wind. A better conductor of heat than either water or snow, it yet provided a protective blanket over the lake. It was far colder above than below. Snow falling more rapidly with each hour lay across the ice, caught the wind and bunched against itself and added to the protective layer. A billion times a billion essentially hexagonal patterns with their beauty lost and forgotten in the very mass of the fall added a quietness to the world. Sounds were absorbed and only the wind and the occasional crack of a breaking limb were heard. But the harmonics were gone, just hard, initial sounds. Snow cleans everything on first fall, ground, air and sound. Below there was just the hiss of the water and the miniature sounds far beneath our range, but critical in the life of a salmon fry. The light level, of course, fell and *Nerka* sank toward deeper places yet.

Above, far beyond *Nerka*'s world but interacting with it as all forces on earth play upon another, falling snow passed through a mass of yet unfrozen water droplets. The crystals picked up a coating of rime and fell to the lake's newly solid surface as tapioca snow. A fog, a sudden and short-lived product of a pocket of warmer air, blew across the lake and the shores and froze midair. Ice needles formed and trees and shrubs took on the familiar look of winter fairy dust.

It was a world transformed. A dropping temperature catching a world of land and water unprepared played trick upon trick, and within thirty-six hours one world gave way to another, another logically next in line. Animals died in those first hours of the Alaskan winter, birds and mammals both. They were the least of their kind or at least the most unlucky. Others survived and steeled themselves against what their instincts knew better than they would come. Those animals equipped to avoid the winter scene quickly adjusted. Grunting bears went to ground with little grace on the north-facing slopes of available hills, the better to be snowed in and protected. Squirrels went to their trees and barely bothered to scold each other, aching as they were with the sudden winter hurt.

Below, *Nerka*'s kind was not unaffected. There was mortality again. Each phase of the salmon's life is designed, apparently, to cull, to clip, to remove those fish furthest from perfection. And fish of other kinds, too, rose to float and bob against the rough underside of the new ice of spicule and needle and rime.

Nerka sank deeper. Adjusting, as he had since the time of his spawning, he automatically sought ways to survive the new and unfamiliar forces. The light was different, the water was different; he had come to a different place. Nevertheless, even as hunters were leaving fresh first tracks in the snow beside the lake the hunters in the lake, too, were aprowl. *Nerka* settled in some weeds near where others of his kind had also instinctively come and within inches of where he hovered for the moment a trout struck. It was a slashing, a twisting, a sliding past and when the trout was gone another four salmon had died. *Nerka* felt nothing for them, for he felt nothing for himself. Unaware of self, aware of neither life nor death—except how to live the one and avoid the other—he could be involved with nothing that could have even the most rudimentary suggestion of emotion. In the presence of death he knew only to move away.

The food supply in the weeks that followed dwindled

but did not vanish. There was enough freely available protein in the lake so that *Nerka* and those others destined to see the ocean in following years could survive. It was a time of diminished growth, a time when the striations on the scales of fish grow smaller, finer and form the clear winter bands. It was winter, a holding time, a time to be seen through, for it held no reward and no promise save that which would surely follow. In a swinging pendulum this was the least time, the slowest time, and in the deep of the lake neither more nor less hazardous than better days. *Nerka*, next to be a smolt, would see it through.

❧ 15

Expanding New Worlds

From *Sockeye*

I said "a puddle or an ocean" and indicated that it made no difference. Of course, that is only when I am comparing land and water. In the watery world itself there are many subdivisions and they make different demands, surely as different as the land places we know.

Nerka the salmon is anadromous and must move between different submerged worlds—a shallow stream, a creek, a lake, a river, an estuary, and the sea, both the continental shelf and the far and the deep. They are different worlds for certain and the writer and the reader must understand the mechanisms that allow Nerka to adapt to these changes, for the new pressures and stresses each place forces onto the fish define the fish itself. Like its counterparts on land, the fish is shaped by opportunities and threats found in the places it occupies.

And then one day the signal came. Within hours salmon smolts by the millions began to move. From the vicinity of the single basin where *Nerka* had been holding, over two and a half million sockeye smolts began the move. They were going to the sea. They spilled out of the lake, they rose up from the basins and the deeper places, they passed millions of others whose time would come another year. They passed them and pushed for the gate, the lake's outlet stream. Predators played along the edges of their unguarded path, for the fish were too intent to be even normally aware. They were answering a call and a pull too profound; it blocked other things out. The edges of their mass were frayed by greedy larger fish, but *Nerka* was deep in the middle of the mass, deep in the heart of the throng of smolts. They were going to the sea, down from the lake, down from bear country and wolf land, down from the hills along a path as old as the species. The movement away from natal water was only less an imperative than that which would one day call the survivors home to die.

The changes that *Nerka* would be called upon to make as he reached the sea were profound. They could not be accomplished in a few hours or even a few days and so they began in the lake before his departure, before the rush to the outlet stream and the descent to the Pacific world beyond the rolling stones and ocean shelf.

Under normal conditions the salt concentrates in a fish's body are greater than in the water around it. This was true of *Nerka* in the Coppertree system and that is why he did not need to imbibe water. There was a natural tendency for the water to flow into his tissues through osmosis, for that is the direction in which water naturally flows, toward the salt concentration. He lost chlorides, of course, through his kidneys and in his feces, but he had chloride-secreting cells to make up for the loss. The new chlorides were distributed throughout his body by his blood. In the sea it would be quite different. The salt concentration outside his body would be much higher and the flow of water would be away from his tissue. Unless he could compensate for that fact he could *dry out* in the middle of the sea. In the sea he had to take water in, to drink and rid himself of excess salts through secreting cells hidden away in his gills. Water passing through was robbed of oxygen and enriched in chlorides. All of this had to be made ready, this chloride turnaround that made him a fish of two worlds. It was not a nicety or casual refinement, it was the difference, for *Nerka*, between living and dying.

The trip from the Coppertree basin deep, where he had wintered and awaited the spring freshet, to the sea was just over eighty miles. His trip would wind him through land no longer used by man, land of bear and fox and deer and the trails of Indians dead of malnutrition, measles, smallpox, syphilis and whiskey. The river itself changed character often. Where beaver had worked there would be broad ponds and strangely the main current of the river would push on across the center of them leaving small eddies to change places with water that had arrived hours before. It was spring and these places were at their deepest. They would be quieter come maximum sun.

At other places the land hunched its shoulders up around the stream, and a cut through harder rocks turned the stream into a race; broken boulders that had caved in from the walls over centuries of frost littered the bed. The smallest crack in the hardest rock can be expanded with the advent of water followed by freezing temperature. It is the microcreep of rock-ruin in a winter land.

In these fast places *Nerka* and the other smolt rode the

race to the next quiet place. They could not swim, they could only point themselves with the current and swirl with it past rocks and submerged debris. The rocks were worn by centuries of pounding water and even when the small fish brushed them they would slide along and escape real injury. Fallen trees, though, were ragged and dangerous, and flotsam caught up in tangles meant for a world of air were traps. Smolt caught in them quickly drowned or were crushed. Luckily *Nerka* knew by instinct to keep to the faster middle water and to avoid the lure of a gentler eddy that could suck or whirl him into an accidental weir.

Raccoons and foxes moved along the banks as the millions of smolt swept past. They pushed their muzzles into the mass of dead and dying fish caught in the tangles. A smolt alone was not worth the hunt but a mass of them pounded into a near-paste by thundering water and the bunching of more dead fish was a treat.

At last the thrust to the sea brought the smolt to a place where the land fell away in steps. One moment swirling forward in an accelerating stream and the next, in groups of hundreds, they were plummeting down seven and eight feet into pools. Pushed to the bottom by the weight of the water that carried them over the edge, they righted themselves and instantly reoriented themselves downstream. Pushing hard, they rose against the pounding above them and rode free on the far side of the pool, only to come again upon a rush that led to another fall and another pool. There were seven waterfalls over which *Nerka* had to plunge and whose slamming effects he had to survive. At last, after two days and nights of constant motion, he reached flatter land and broader, quieter waters. It was here he began to taste the first real wisps of chemical change.

He had come to the place where the push of the ocean tide could flavor the river. The chemical adjustment in him began accelerating. He had very little time left before he had to become a fish of bitter water in a world as different from the Coppertree deeps as, perhaps, the sky is from the land.

A strange element in the life that *Nerka* had led so far was its absolute solitude. As an egg he had been hidden in gravel millimeters away from the others of his kind. As an alevin, too, his closeness to other sockeye was the thickness of a few sheets of paper. As a fry in the lake he had lived with millions of his kind and his paths had intersected those of millions more. In his descent through the maelstrom of the lower Coppertree in flood his adventures as a smolt had been shared with hundreds of thousands of young fish experiencing exactly what he was experiencing, undergoing the same tests. Their bodies brushed, they were hurled together, they fed together and very often, by the scores of thousands, they died together. In every way they were alike, yet each was as alone as if light-years and not water molecules separated them. They actually shared nothing and never acknowledged each other. A fish does not know its identity and therefore cannot truly know the identity of another of its kind until it is time to breed; then smell overwhelms it and directs all of its actions and all of its energies.

It is true, of course, that different fish have different patterns, different ways in which they move and orient themselves to essential goals. Fish can align themselves so that schools of one kind can pass through and around schools of others. But it is wholly an unconscious thing, for a fish can never belong in a conscious way. A fish cannot learn from another fish any more than it can really learn very much from its own experiences. It may respond as a result of deeply traced response patterns, surely there are negative and positive incentives, but here again it is a question of reflex. A fish's reflexes learn what the fish itself can never know.

It was alone then that *Nerka* descended to the sea. Until the act of mating occurred he would be a single dot of light in the Milky Way denying all other stars. He cared for his fellow fish as a star cares for a star and as the sparks relate to each other in a comet's ice-fire tail. *Nerka*, amid millions of his kind and thrust rudely into the world of billions of other creatures who could never taste sweet water and survive, would be alone. He read other sockeye as he read all that went on around him, but it was without knowledge of life, without the first inkling of concern or any power to know. All that he was had come to him at his hatching and all that he would ever be would wait until the hour when he again reached the redd where as an egg he had been fertilized and his chemical memory had begun. He had come to the sea with the single purpose of maturing—he would grow and mature and then, when ripeness again sent demanding messages throughout his glandular system, he would smell and taste his way home and use other cues as well. The spark was coming to the sea for it was his time to do so, his cosmic, eternal place in the cycle.

.   .   .

The attrition along the route from Coppertree Lake to the last tidal reach of the Lower Coppertree River had been constant. There were fewer smolt now, fewer by many, many thousands. Most of the smolt didn't stop to feed once

they started their migration downhill to the sea. They relaxed their usually alert defenses and were preyed upon all
along the way. In fast streams fast cousins, trout, slashed
through them and large gulping carpish fish swallowed
among them with glassy-eyed indifference. Natural hazards chewed away at their numbers and only the best made
it, the best and the luckiest.

The smell and the taste of the sea were the first clues
Nerka received. Bitter threads of salinity worked back up
into the river and it was along these ribbons of future and
remembered taste that the young smolt moved toward their
destiny. The sea taste became stronger by the hour; too,
there was a two-way movement of water. It pushed inexorably toward the sea, but a second and equally inexorable
force pushed back in an unremitting cycle of tidal activity.
Water flowed across water in turbulent layers and the smolt
had their first meeting with the moon.

For several hours *Nerka* held back while new sounds and
sensations reached him in increasing crescendos. He circled and listened with his hidden ears and his lateral lines.
He moved from place to place constantly reorienting himself. He was in a small tidal place, a small cut in the river's
bank that formed a slow eddy and a quiet place to hold. His
body was making the final adjustments and he was sufficiently keyed by the quality of the new water to automatically allow these adjustments time to occur. He was repac-
ing his own body processes. Then it was dawn and *Nerka*
moved out of the eddy and back into the main stream. The
river broadened rapidly and surged with the sea beyond.
Overhead gulls by the thousands gave rude noise to the
sky, but *Nerka* was deep enough not to attract their attention. Smolt that somehow found themselves too close to

the choppy river surface were often culled. It was quick and easy—suddenly a massive disturbance exploded through the surface and they were gone. Bubbles sank and then rose, but the young fish were already airborne in the gut of a bird no less rude now than before. Sometimes a gull would drop its prize and greedy warriors cut through the sky. The fish would become a toy, a thing to grab, to bounce and then to swallow.

The sounds were strange to *Nerka*, the pressures that played upon him in concert and fury. The thumping beat of a tanker off to the side, the thud-crush of a pile driver sinking the footings for a new pier at river edge. Beyond the mouth of the river, as *Nerka* approached, there was a gradient over which an incoming surf eternally rolled. Bottom debris of shell and pebble and rock tumbled and surged, slushed and sighed and rattled. It was all new and in some strange way it excited *Nerka*. He pushed faster into the bitterness of the sea. Tasting his last truly sweet water for several years, he drove on directly ahead as the bottom fell away from him. He was in deeper water than he had ever experienced before.

As he pushed to sea he entered the Alaskan Coastal Domain. It would take him days to cross the area into the gyre to the south of it. He would then come to the vortex known as the Alaskan Gyral, and in that gentle elliptical vortex he would sample the truth of the sea and the promise of his inheritance. He was no longer a lake fish or a river fish. He had been transformed. He was an ocean fish, a marine animal, and all his chemistry was now completed. In those first hours he began to dehydrate as chlorides were sucked from his tissues, but he had been well prepared and before any appreciable harm was done he reversed the trend. He was drinking seawater and excreting salt beneath the opercula of his gills. His system completed the adjustments in time and this hazard, too, *Nerka* survived. Not all smolt do.

The waters of the Alaskan Coastal Domain drift westward along the south coast of the Alaskan Peninsula at a speed of eight to ten miles per day. They move along the top of the elliptical vortex that ends where the final Aleutian Islands stand in solitude among gull cry, sea mist and bleached grass. Here, too, are the shared memories of broken boats and broken men. Much of that water then swings north and east and circles the Bering Sea in a counterclockwise movement at a speed seldom exceeding two miles per day. It is a complicated system with some water from the Bering Sea moving almost due south at its western end in the East Kamchatka Current. Some of it moves eastward again along the southern edge of the Alaskan Gyral until it comes up against the coast of Canada and the United States. Through all of these currents and in all of these worlds the salmon move during their years at sea. As they sweep past coasts and islands they sense sweet water from Asia and North America, from the Arctic and from temperate zones, but only as fragments, only as highly dilute hints. None of this attracts the fish to shore until they are ripe with their time and then only that special sweet water in which they grew. As thin as those dilute fragments of river systems may be, in the sea they are distinguishable to the salmon.

The top thirty feet of ocean held the most promise for *Nerka*. It was summer and that upper zone was homogeneous, warmer and lower in salinity. The temperature ranged between 50 and 54 degrees. A hundred feet down the ocean was even more constant—it neither warmed nor cooled and the salinity was almost 2 percent greater.

Nerka shared his world now with a greater variety of animals than ever before. His smallness was an advantage, for many of the great animals that hunted in the coastal waters off Alaska will only seek more satisfying prey. In time he would interest them as well, but as a smolt new to the sea he escaped their notice or at least did not sufficiently arouse them to trigger an attack. At one point the sea around him seemed to explode. A wave struck him that sent him spinning off in uncontrol. Over and over he flopped although he fought to right himself. His whole world—sight, touch, pressure, taste, scent—was full of whale. *Orca*, a thirty-foot-long bull killer whale, cavorting with a pod of his kind had flopped free of the ocean, gone up into the air and then come down with tons of pressure. It was this that had caught *Nerka*, flipped him and nearly crushed him. Then up again, up came the killer whale pushing a wake ahead that exploded with bubbles. Black and white and uncounted thousands of times *Nerka*'s size, the whale rose and again exploded past, sending *Nerka* spinning away a second time. For several minutes the bubbles continued to rise as a silvery veil that lessened and slackened until the sea below returned to calm. The pod of whales was miles away by then doing something that *Nerka* could never do. They were playing, they were calling to each other, they were relating to each other and enjoying the sensation of life. They were interacting and even, perhaps, showing off. The salmon smolt and the killer whale were both sea animals but the one, the greater in size, was ahead in evolutionary gifts and skills. The history of *Nerka*'s kind had gone off on a different spur long, long before the killer whale's ancestral forms had even reached the land, much less returned to sea. But now it was back. As a mammal it was up from the reptile, in turn up from the oldest form, the fish. The killer whale, child of the fish, had some ancient ancestry in common with *Nerka*. It is a line that can be roughly drawn. But how different the form now, on encounter, how unequal the two perfections.

16

A Secret Journey to the Sea

From
Sockeye

Once truly in the sea, for one to seven years, the salmon meets entirely new threats and chances for survival. As it moves deeper and farther away from the continent to which it must one day return, the fish takes on new size, strength, endurance, and skills. Blindly responding to chance and a complex hereditary package we can barely comprehend much less explain, the salmon becomes a part of its strange new world, and the reader, more a stranger even than the salmon to the surge and cold of deep sea places, must go along.

The writer cannot pass a reader along from place to place without a strong bridge to carry the weight of curiosity and even concern. That bridge is the animal whose life is being traced. We go with the fish from place to place in time and space just as we rode with the condor, stalked with the cat, and bumped and bungled with the bear. Always, though, we have to care. The more we care the further we are willing to go and the more about each place we are willing to learn.

No one has ever really traced the course of a sockeye salmon at sea. That is a secret thing etched in deep and secret places. It is likely that no two salmon ever live exactly the same life and that no two courses are ever repeated, although even the infinity of chance would seem to be overwhelmed by the numbers of fish that come down from all the freshwater systems that feed the northern Pacific world.

Nerka slipped westward early in his ocean career, then south at the western end of the gyre and then east with the vortex path. He was near the coast of the State of Washington, following a rising bottom and feeding well on a rich harvest of tiny ocean life. The water was cold and cold water sinks. Good vertical currents kept the layers that lay upon each other from air to seabed richly fed with minerals. Life grows well in mineral-rich waters and life upon life feeds in a slowly ascending scale.

At one point *Nerka* came upon a broken rocky place and drifted among former mountains and over rents and fractures of ten-thousand-year-old quakes and tremors. Some of the spaces between the enormous boulders had filled with sand. It was sand made there of disintegrating rock, for the tide and surf were strong, and sand carried to sea suspended in river water at flood. There were other shapes here, too, boats and ships tried too hard by weather. And on the decks and along the rusting plates of these artifacts of an inhabited coast the sand lay, too. Lines of rivets pushed up under the sand, some still barely showing, marching off into the murk like impossibly oriented mountains in miniature. It was a place of graves and abruptly ended history, a place for fish now, fish, their enemies and their prey.

There was a slushing sound, for the water was shallow enough and the surf above was felt below. The whole world rocked as it moved first in and then out, great surges of energy with a vast ocean behind them moving in against a sloping floor that led up, up to a continent beyond the furthest edge of the submerged shelf. From the distant gloom *Nerka* drifted into view. He turned one way and then another, a fish among many, of many kinds. He wasn't feeding at the moment, having recently found and taken the smolt of a different species, a small fish he was able to engulf. He rode the submarine swelling and let it lift him closer and closer to the battered hulk of a dead ship. Inside, the surfsurge rode through a narrow opening in the hull and played against a door that had been left ajar as the rammed and battered ship had overturned and died long ago. The door, rusty and growing thinner each hour of each year from the erosion of salt water, banged in a rhythm of the ocean push. It was a dull thud without echo, but it rang hollow and beat a tattoo that carried far and was heard by millions of creatures. And millions had come to investigate it. Predators had learned that, in time, without ever understanding it, of course, it was a place to prosper in. Sharks came here to feed and other creatures as well. The thunk of rusty metal on rusty metal carried far away into the sea. It was an unfamiliar sound.

Nerka drifted closer and closer. As he passed across the side of the ship corpse barely inches above the plates, the swirling water hissed sand grain against sand grain and the shell and pebble debris that inevitably covers every horizontal surface in the sea. There was the slushing of the water itself, then its play on metal, the thud of the door within the hulk and the hiss of gravel. But then, suddenly, there was another sound. No man could have heard it, but a salmon could. It was as foreign to *Nerka* as the metallic rattle and groan of the dead ship. He could not locate it except that it seemed to rise up toward him. He held and turned first one way and then another, trying to obtain maximum purchase on the sound, using the tubes of his lateral lines and his hidden salmon ears. In fact, behind a tangled piece of hull not a dozen feet away a crab on extended legs moved sideways in search of food. He, though, was hunted even as he sought prey. An octopus, smartest and second-largest of the invertebrates (squid are bigger), pulled back and changed color until it was itself peeled paint and rusted metal. It pulled back, and the convulsions of its eight arms created a series of signals new to *Nerka* and one, indeed, that he might only rarely sense again.

Somehow the crab too awoke to danger and stalk-equipped eyes rotated in sweeping assessment of the immediate and always dangerous world. It froze in place, but it was too close and too late. The octopus, a monster fourteen feet across its outstretched arms, rotated its body until its beak pointed in the direction of the crab. The octopus shot out from its hiding place, snatched the crab and pulled it in toward a snapping, black parrot's beak. While alive the crab was pulled apart. The octopus did not need to envenomate its prey although it could have spit its deadly chemical down into the open wounds. Its beak stilled the crab forever.

The movements made by the crab and the octopus arrested *Nerka*'s forward movement. Repelled by the strange but clearly threatening movements and sounds, *Nerka* turned away quickly and vanished behind a section of up-folded iron plate. Sun played through rivet holes and lay dappling across the sand and shell beyond. *Nerka*'s sudden movement thwarted death again, for a very young shark on coastal patrol had dropped down from above and spotted the young salmon just before he moved. The shark fell upon an empty place and turned away in time to take another fish of *Nerka*'s kind who was just moving into view. It was an older fish and the shark bit it in half. The head was severed and sank to the seabed below where crabs who had not yet met their octopus would feed upon it. There was no one to observe the interesting fact that the face of a dead fish is no different, really, from the face of one that lives.

Nerka began his drift to the north again, beyond Washington State and the United States, along the fog-bound and high-treed coast of British Columbia. He drifted and grew, gained in size and strength and came each moment closer to the distant goal of maturity. A sea creature full and sure, he explored his world for hundreds of miles without curiosity or concern, drifting, marking the hours and the years with growth. His mission was to become complete and full of the nature of his kind so that he could make his deposit on the future and then die. His role, like the role of everything that lives, was immortality. He grew toward that, he escaped hazard almost every hour of every day, and that, too, was his gift to the future. In the sea his course was seemingly random; in time it was inexorably set and true.

·   ·   ·

As *Nerka* grew, so did his power as a swimming animal. It is almost axiomatic that the larger a fish is the faster it can swim—and with each inch of growth *Nerka*'s powers increased.

Muscle bundles lay along his body from head to tail whose full purpose was propulsion. He flexed sideways much more readily than he did up and down and was able to ripple the muscles along both sides in perfect coordination. A bundle contracting on one side would be matched by its exact counterpart expanding on the other. With body curve and rippling muscles along the length of his body, *Nerka* moved forward and ended each ripple with a thrust of his tail. That too drove him forward in the perfect symmetry of an animal in harmony with its medium. His streamlined shape allowed the water that naturally piled up in front of his head to slip free and flow along his sides as his muscles followed their rippling sequence. The flow added to his drive.

His tail, a fan-blade of thin bone rods covered by a stretched membrane, was attached to the end of his spine by a series of bony plates, muscles and controlling ligaments. It was as superbly a coordinated member as his senses and his lateral muscle bundles. It waved easily from side to side, and was always ready for an extra thrust toward life when death came close and offered itself for chance.

There is a kind of rule of thumb men use when they think of fish and wonder about them in the sea. A fish's sustainable speed is about seven miles an hour for every foot of body length. Men like such rules even though there are often more exceptions than creatures willing to submit to them. In the salmon, though, it is not far from true. There is another rule as well. A fish should be able to add 50 percent again as much in a burst of speed. A one-foot fish, then, should "cruise" at from six to eight miles an hour and explode away from danger or come down on prey at from nine to twelve miles an hour. Salmon are faster than many fish; and *Nerka*, although still far from a foot in length, was yet nearing this power and this speed.

When *Nerka* moved quickly, for whatever purpose, he was called upon to make constant adjustments in his own chemistry. If he swam a normal easy course, moving without urgency from place to secret place, the metabolic wastes he built were removed by his bloodstream and passed away without special compensation. It was a chemical rhythm added to the mechanical, and together they constituted the system of his fish life.

But sudden bursts of energy, an explosive hunt or escape tactic, overrode such easy ways and a potentially deadly poison quickly began to accumulate in his tissues. After exertion *Nerka* would rest and allow the lactic acid to bleed away from his muscles and be carried off for disposal. As long as he lived, in the sea and in his eventual uphill fight to the far reaches of the Coppertree, he would balance exertion with rest and quiet his chemistry before exerting himself again. As a potentially highly active fish, he was more susceptible to this hazard than were many other species.

Nerka was equipped in another way for a highly active life. Like a number of fish he had a swim bladder which he could inflate or deplete to help control his buoyancy. Many fish, most perhaps, have swim bladders, tough little sacs deep inside of them, at their precise center of gravity. This provides them with the easiest means of trim; the least amount of work is thereby needed to remain level in the water. *Nerka* and his kind, though, have sacrificed that ease for a tactical advantage worthy of only the most active fish. *Nerka*'s swim bladder was below his center of gravity.

When hovering he tended to be rolled up sideways, but he was muscular enough to compensate for that. When maneuvering in a current or when taking prey near the surface—or simply on a rising thrust—the off-center placement of this buoyancy compensator gave him great advantage. He could turn faster and harder than other fish, for on a bank the position of the sac tended to roll him on his side. Without extra effort he was in a position to use his forward motion to carry him around the curve, be it horizontally or vertically oriented, and on from there. If he broke to the surface, he could make an effortless 90-degree roll onto his side and in a single slashing movement go first up and then down. The arcs he could describe in the sea and would later use in countering river currents were things of swift beauty, the awesome purity of a life perfected by time. He was an athlete among fish, a lithe creature in the sea—for this he had been designed.

Nerka often traveled in company with others of his kind. They did not relate but sought the same conditions. They cruised at a depth of one hundred twenty feet although they swam more often in the last forty feet of ocean before the air began. They tended to concentrate in the subarctic zone of the northern Pacific, between forty and sixty degrees of north latitude. *Nerka*'s rate of travel varied; there were days when he moved no farther than fourteen miles in twenty-four hours and other times when he would travel as far as fifty. It would be several years before he would have anywhere to go except toward the next thing to stimulate or attract him, or away from the next set of stimuli to repel him. He was biding nature's time, preparing for his role.

The food *Nerka* sought, allowing for the differences between his two worlds, was not unlike that which he had hunted in the Coppertree system. Pelagic, planktonic co-pepods, amphipods and other minuted invertebrates that drifted like clouds in nutrient-rich seas. He took young fish as well, herring at times, and other species. Larval crabs were a food often consumed and probably taken whenever encountered, and also the smallest of squid. As he grew he would take larger squid and larger fish, but always the crustaceans, the drifting clouds of animals, would be favored. He did not often rise wholly to the surface, but even when he fed just below it, his body roll would sometimes send his tail rippling half in air before he arced again into the ocean dark. There were times when he seemed to do things that had no real purpose and one is tempted to say he played, that he created pleasure and perhaps that complex affair we call fun. But we are not certain that this can be allowed a salmon.

He was becoming faster and more greedy. He fed constantly and he grew. He pursued food and smelled out waters where upwelling currents kept them rich and full of crustacea.

One day, if he survived to maturity, *Nerka* would be slightly over two feet long. A number of factors would influence not only the rate of his growth but his maximum size. The amount of food he consumed was paramount, but so was water temperature. It is believed that warmer waters foster greater growth, but many of the factors at work there are still secrets the salmon keep to themselves. There is a periodicity to the growth of *Nerka*'s kind, too, for salmon of one year seem to achieve greater size than salmon of another, even from the same river, perhaps from the same gravel beds and certainly as tenants of the self-same sea.

Maturity, too, a mystery of salmon life, when would it come? Heredity might be a factor, rate of growth another, or again a strange pulsing. On land the populations of animals vary markedly from year to year, and some see this as linked to cosmic happenings. If the salmon, too, are linked to events on distant bodies in the sky, they have kept that secret and we still search among those fragments of their lives we know for answers that only possibly are there. Some salmon mature not much after their second year and others when they are eight. It is at least strange, for they can be of the same species. And so *Nerka*, with all of these secrets inside him, swam north and then west again passing through galaxies of planktonic animals and consuming them as he moved, and through a world of shimmering bubbles, shards of sun diffused in a world of water, a world of life and a world of death. Trimming his buoyancy, moving vertically and horizontally through infinitely variable choices, bunching and relaxing muscle bundles, aiming himself, a silver dart, he swam the Pacific world, a secret animal in many a secret place.

❦ 17

Before the Forest There Is a Tree

From The Forest

We have read a great deal about animals so far in this collection with plants playing no more than supporting roles. That is natural for we are animals ourselves and unabashedly anthropocentric. It is easier to relate to a bear or a mountain lion, even to a vulture or a fish, than it is to a milkweed plant or a growth of poison sumac.

Plants, though, are organisms that must "learn" (read as adjust) to adapt or die. The weather dictates most of what happens to them, but they have their predators, often their protectors, and are very like animals in a great many ways. Of course, plants as individuals tend to be less mobile than most animals we know outside of the sea, and most animals are more intelligent than most plants. We accept that to be generally true.

In this first reading from a book called The Forest, we meet one of our leading characters, a hemlock tree. It is as surely a part of the place as are any of the animals that depend on it, attack it, or benefit from the chemistry it builds, stores, and will ultimately return to the soil that it creates even as it grows in it. If readers try

hard enough and are open-minded enough, they can care about a
tree, deeply and without shame. The author of this book cares, I
know, and at least some effort is made here to make others care as
well. If a natural history writer can't get people to care, then
perhaps he or she should write auto repair manuals or gothic
novels.

The hemlock on which the eagle had landed was more than 450 years old and stood 200 feet tall. Its kind came into being before even the dinosaurs walked the earth—needle bearers to fill the first dryland forests—millions of years before the oaks and elms, the walnuts and aspens, and other more complex trees came into being. Grass came later, too, and so did wild flowers and all flowering plants.

The last glacial period, between ten and twelve thousand years ago, was unkind to the hemlocks of North America. They existed through that period, true enough, and survived, but in smaller numbers, in fewer places, and perhaps in smaller sizes. About nine thousand years ago, in a kinder era known as the Pre-Boreal, things were better suited to the hemlock kind. By the time the Boreal period of 6000 B.C. was upon their land the hemlocks were in command of their part of the woodlands. And although the signs are small and slow in revealing themselves, the weather has cooled since then, since about 4000 B.C., and it is possible that the hemlock hold has slipped accordingly. Such changes are not directly visible to an unaided eye with a limited span of years, and only the count of seed and pollen left unrealized in soil deposits can prove or disprove such theories. But the hemlock still flourishes and enriches a world that must in turn be rich for the tree to thrive.

Hemlocks are simple trees. The cones that protect the seeds and the needles that are their true leaves are of a primitive but highly successful design. The cones exist for the single purpose of holding the seeds of the future until it is time for them to be dispersed. They are tough, those cones, with each scale harboring two unshielded ⅛-inch-long seeds at its base. There is no anther, no stigma, no ovary—all refined parts belonging to younger species. The hemlock needle is dark green, shiny, and grooved above. Straight and plain, it is protected against weather changes by a coating of wax. Resin helps hold off the effects of decay. The roots of the tree run deep and are more simply designed than those of species that came later. The hemlock is an ultimate tree, a tree unadorned by climbers or runners or bushy growth.

From Alaska to California, discontinuously, the flexibly tipped western hemlock shares the great upland forests with the Douglas fir, the Sitka spruce (lower down), western red cedar, alpine fir, red fir and white. Massive and bothered by few insects, these conifers preside over a complex web of animal and plant life and some forms in between. Spawns of a temperate maritime climate of great stability, the trees stand for centuries, in some species for millennia. As they reach great age, their needle weight increases, and the massive branches interlock, sealing off the forest floor below from the rays of the sun that have, up to then, shone through like the light shafts in a cathedral. Eventually, though, the mass of the needles becomes so great that the trees prune themselves, as branches crash down to add to the debris on the forest floor and allow smaller, younger trees their time in the sun. It is then that the western hemlock flourishes in a world of giants. Along with the silver fir and some of the cedars, it grows toward the sky.

Very often younger trees grow up from patriarchs that have been toppled by wind and rot. The ancient trees down in the evolving humus become the nurse logs upon which the future grows. That was how this tree, the hemlock on which the eagle rested, began. Almost half a millennium ago, a cone on its parent tree had matured and opened, releasing the seeds, which then fell to the ground. One seed had been blown sideways by the wind and had landed on an older giant that had died and begun to decay. Falling into a small crevice of this nurse tree, the seed had waited. Inside the seed an embryo tree lay, with all its leaves, its stem, its critical root point. Rain and sun came in proper sequence and intervals. The needs of a new generation were met.

The embryo began to grow. There was enough warmth, enough moisture and nutrients surrounding it both within and without the seedcase for its growth and survival. The growing embryo split the seed, and the tough root point emerged. It bent over and began to grow down, invading the decaying fibers of the nurse tree. Soon the tiny point was drawing both food and water from the old tree and the soil it was rapidly joining. Needle-shaped leaves pulled free from the seedcase as the tree passed its next critical test. It began manufacturing its own food. Root hairs had formed deep down where the root point led, and the terminal bud below the empty, unattached seedcase lay beside an infant tree on its way to the sky.

All things on earth are finite. All the needles in all the coniferous forests are measurable. If they seem without bounds, it is our inability to comprehend them that is at fault. Our only encounter with the unmeasurable occurs once our thoughts leave this planet. Infinity is probably a temporary condition, however. As the lifespan of stars and species goes, it may one day be a classroom toy for slow learners.

Among the finite substances on and around our planet the masses of which we guess at still, rather than truly weigh, are the gases oxygen and carbon dioxide. Life on earth is wholly dependent on them. We believe now, although our scales and yardsticks are not as precise in the great affairs of our environment as we would like, that the carbon dioxide in our atmosphere is renewed once every three hundred years. Using the same primitive tools, we have come to believe that the free oxygen in our planet's system is renewed every two thousand years. These facts, whatever their actual dimensions, tie us to the tree and the tree to the eagle. We all are locked into a web of exchange.

An estimate generally accepted today places the amount of carbon dioxide processed annually on earth at 200 billion tons. Of that conversion, probably 80 percent occurs in the upper layers of the ocean where an unimaginable, if entirely measurable, number of minute plants live. The other 20 percent, still a critically high percentage, apparently occurs in the gardens of the world, both the gardens of man and the gardens of nature, such as the forest in which the eagle sat.

This processing of carbon dioxide, known as photosynthesis, was operating within the infant western hemlock from its first exposure to the sun. Light falling on its needles and leaves was absorbed as energy. Free energy, always available in every daylit hour, fed the forest through each of the green plants growing there. The work was done in each by a pigment in the leaves called chlorophyll. That was the principal pigment at least, but there were, in fact, others working at other places in the spectrum.

In a miraculous link to the cosmos, light supplied the energy required by each plant to produce the organic molecules which compose all living things. Each green plant, green because of chlorophyll, was able to extract the free carbon dioxide surrounding it in the air and cause it to react with water from the air and soil and form organic compounds. It stored chemical energy for growing cells. As this occurred, oxygen was released into the atmosphere as a kind of exhaust fume. Animal life in great part sustains its cells with the waste products of plants. If the hemlock could have absorbed light of only one wavelength, however, a large part of the sunlight that plowed into the forest's depths from distant space would have been irretrievably lost. Although chlorophyll could work alone, could transfer carbon dioxide and water and light into the hard substances of life, photosynthesis is most efficient when two or more pigments work together. Light in shorter wavelengths falling on the hemlock, lost on chlorophyll, was trapped by its companion pigments. When the reddish beta-carotene rejected the red or long wavelengths—that is, left them for chlorophyll to deal with—almost the entire visible spectrum was put to use by the tree.

The state of energy conversion was stable, long-lived, and uninterrupted and would remain that way as long as that energy was available to the forest. While photosynthesis was proceeding rapidly in one cell of the hemlock, that cell had the ability to store light energy or pass it along to neighboring chlorophyll molecules for immediate or future use. The sun's energy was shunted from cell to cell within each needle on the tree, and energy that might have been lost as heat became the stuff of photosynthesis where and when it was needed.

The needle-shaped leaves of the hemlock no less than the eagle were bathed by energy born in the nuclear holocaust we call the sun. That sun, although 93 million miles away from the forest (that is, 270 times closer than Proxima Centauri, the next nearest star), was as much a part of the forest and the life cycle of the eagle and the hemlock as any object or force close enough for either to touch. The eagle and the tree were of Earth, and Earth is of a system tied to Sol, our sun, and Sol contains 99.9 percent of all the mass of our entire system.

The continuous nuclear explosion that is a star came to earth and visited the eagle and the hemlock with energy made benevolent by the length of its journey. Their kinds were born of that energy and forever linked to it. The human co-planeteers of the eagle and the tree can never forget that their subjects are, like themselves, linked to a cosmos. That is the larger truth of the tree and the eagle, their link with everything that has ever been, everything that is, and everything that can ever be. It is in this concept that so many people see a master plan. If the plan does in fact exist, the assault by some upon the tree, upon the bird, and upon man, who alone on earth can think of these things, is most difficult to comprehend. He who would strike the bird or the tree or man strikes at a child of the sun, and the anger of that parent is beyond the farthest edge of the human mind.

The tree was a perfect combination of elements, each of them reflecting the years of the giant's growing. In the center lay the heartwood. It was the supporting column of the mature tree. Heartwood is dead, it cannot grow, but it will not decay or fail the tree by losing strength as long as the rest of the tree remains intact and in balance with its surroundings. Girding the central column was the sapwood. Through it ran the pipelines that carried water from the roots to the leaves. As the inner layers of sapwood lose their vitality, they join the center column as heartwood. Surrounding the sapwood was a thin but vital layer known as the cambium. Each year, stimulated by auxins, or hormones, this layer produces both new bark and new sapwood on its outer and inner surfaces. Lying outside the cambium layer was the part of the tree known as the phloem. Just as the sapwood carried water from the roots up to the leaves, the phloem carried food down from the leaves to the rest of the tree. As growth continued within the cambium layer, the phloem was pushed out to become true bark. And beyond the phloem was that bark, the means by which the tree protected itself against heat and cold and some enemies.

These transitions, sapwood to heartwood, cambium layer to sapwood and phloem, and phloem to bark will continue as long as the tree is alive. When they stop, the tree will die. Its heartwood will fail soon after that, and all the chemicals the tree contains will circulate again as the tree becomes the victim of weather, fire, insects, and time. It is the chemical harmony of the forest, and if anything approaches perfection on this planet, it is that.

Even as the eagle rested, looking out across the valley floor, a war was being waged within and on the tree itself. Tree-killing bark beetles of the family Scolytidae had invaded the forest some weeks before. Working first on fallen logs, then attacking standing timber as their numbers increased, they ground and crunched their way toward the living heart of the forest. Small, dark, compact, these insects had found the tree of the eagle's perch several days earlier.

They were still trying to make inroads. The bark of the western hemlock is between an inch and an inch and a half thick. It is deeply seamed and has broad, flat ridges with close, thin, cinnamon-brown scales. It was through that bark the tree killers had to bore to find the pale yellowish sapwood beneath. There, where bark and wood met, the bark beetle could excavate its tunnels and galleries and plant its eggs, from which tree-eating larvae would hatch to further the march of destruction.

When the first beetles reached the wood of the eagle's hemlock, they found a tree in water balance; within its tissues there was enough moisture, but not too much. As they began scoring the cells of the wood itself, they encountered richly laden resin cells, which exploded and killed many of the beetles before real harm could be done. Had the tree been badly stressed by drought, as can sometimes happen on western slopes, the resin would not have jetted out, and the insects could have continued their assault unimpeded. It could have meant the death of a tree. In trees like the hemlock a water shortage is felt in many ways.

The hemlock had outside allies in its war against the invading tree killers. The Cleridae, checkered beetles, voracious predators faster than the bark borers, had swarmed into the area when the bark beetles first began settling in. The female bark beetles had sent out chemical messages, pheromones intended to attract males from miles around, but the checkered hunters had detected them, too, and responded on a different kind of errand. They pounced upon all the exposed bark beetles they could find and devoured them. Then, where they could find holes in the bark, either they pursued the borers into their lairs and ate them there, or they deposited their eggs. Soon after those eggs hatched, the predatory larvae would follow the tunnels of the boring beetle until they came to their larvae. They would feed upon them, larvae upon larvae, related as predator and prey just like the adult forms of the two species.

Delicate brownish black braconid wasps also joined the battle. By means not really understood, these wasps follow the bark beetle infestations from forest to forest and move in at just the right time. Using complex infrared receptors to detect heat, they locate the larvae of the bark beetle behind the inch or more of thick, pulpy bark screen shielding them and drill through with their ovipositors. Thrusting their muscular abdomens downward, the wasps inject their eggs into the bodies of the bark beetle larvae. The eggs hatch there, and the emerging wasp larvae feed on their hosts. The bark beetles, once they have been injected with wasp eggs, are doomed. No force on earth can save them.

Met, then, by exploding resin cells not made flaccid by drought, attacked by hunting clerid beetles, injected with hosts of braconid wasp eggs, the invasion of bark beetles was slowing. It was a momentary setback, though, for these beetles, like all primitive life-forms, are tenacious. The bark beetle has both time and numbers on its side. One day, in this forest as in all others, harm would be done, and trees would fall, victims of small dark insects, any one of which weighs no more than a needle or two from the hemlock's smallest, farthest arms.

18

A Library
of Chemicals

From
The Forest

Some natural history writers relish the fact that they are in a very real sense educators. Others deny it, equating, perhaps, education with very heady stuff and thinking of themselves rather as entertainers or, at most, sounders of alarms. I like to think that we are all of those things and perhaps a few more.

However we may fancy ourselves there is an undeniable fact that in school irregular French verbs, calculus, and chemistry are the three killers that send more people screaming out of school and into the streets than almost any other subjects. You can study nature without irregular French verbs; I don't know a thing about calculus and hope to keep it that way; but chemistry is unavoidable. Most people are so shell-shocked from their school experiences that they will clamp shut like a vault door at anything that even sounds like it might be chemistry. Nature writers trying to describe the places where and the means by which animals and plants live simply have to slip chemistry in in a way that readers will tolerate. It really is basic information and can't be avoided. It is a trick to make it interesting and an art to make it fun.

The chemistry of a forest is a complex matrix of elements, compounds, and time. Organic matter is produced so that more organic matter of different kinds may flourish. To the forest a tree is a stage, a platform, a means of holding critical chemicals in storage and of converting others. No tree lives forever, and all chemicals return to the original mash that means continuing life in the forest.

Rocks and soil and weathering in the forest itself, or high enough above it to wash down into its soil, produce calcium, potassium, magnesium, and phosphorus. Nitrogen is needed as well, but nitrogen comes from other sources, from dust and rain, even lightning. Blue-green lichens are nitrogen *fixers*. They rob the air of nitrogen and work on its chemistry in subtle ways, until the raw nitrogen is fixed into new organic compounds which reach the forest floor when rain falls and filters through the minute lichen forest. The water carries the critical compound with it as it penetrates, percolating through the forest floor debris. When a tree itself falls, it carries its lichen burden with it, and animals that eat lichens leave nitrogen-rich droppings and eventually die and decompose or are in turn eaten and become the metabolic slag of other species.

At least 150 species of lichens inhabit the fir forests where hemlocks grow. Some species produce as much as 15 pounds of nitrogen per acre every year. Every animal and plant in the forest community, whatever its size, whatever its role, must have nitrogen, and a major source for all are the lowly lichens growing along trunks and branches, minute manufacturing centers that draw their raw materials from the air itself.

Lichens are among the least dramatic and most important life-forms on earth. They are the only plants* that can grow on otherwise barren rocks and are, in fact, one of the principal means by which rocks are decayed and turned into soil. Though edible by man, lichens taste quite bitter and are unimportant as a human food source. They do have some limited economic value in short-term projects; some lichens soaked in human urine create the dyes used in all authentic Harris tweeds, and the laboratory paper known as litmus, used in tests for acidity, comes from an extract of lichens. As so-called reindeer moss and in other forms, lichens are vital to animals both large and small.

It is difficult to imagine what the world would be like had not this association come into being. Each lichen species is the combination of two other forms, a fungus and an alga. The lichen itself is neither an alga nor a fungus, but rather both constituents locked together in an inseparable relationship.

Fungi, those relatively low forms of life that include the mushrooms, can either live free or combine in such obligate relationships with algae. For more than 350 million years, fungi have lived in a myriad of relationships, and one of the strangest is the form we call the lichen, the nitrogen-fixer.

The fungus partner in most lichen species (but not all) determines the size the lichen will take and forms the bulk of the associated structure. Mycelium, for instance, is fungus material in the form of elongated tubular cells designed to invade and break down rock, wood, and bark and turn the substrate into food. The rate at which some such fun-

*Many scientists today place lichens and fungi in a kingdom of their own, no longer calling them plants at all.

gus species produce these cells is astounding. These irresistible penetrating seekers of food may grow as rapidly as $1/8,000$ of an inch per minute. Each cell of the mycelium consists of a wall of chitin—not unlike the exoskeleton of insects—that contains the living material of the plant itself: the protoplasm and one or more nuclei to signal and control the unit. It is the whole of the forest reduced to microscopic size, the endless repetition getting smaller and smaller until it drops below the range of human vision.

Once a fungus member of lichen relationship has colonized the bark of a large tree, its hyphal tubes begin their spreading invasion. Soon a network has formed within which semigelatin motes of algae are enmeshed. The fungi network and the algae prisoners together constitute what we call lichen. That is the association arrived at so many millions of years ago, which allows many lichens to fix nitrogen and give it up to leaching rain and dew to feed the soil in which the hemlock grows. These life-forms use the tree as a platform and feed in part from its bark. The bark, in turn, feeds the roots of that host tree and all other life in the forest. Thus hemlock and fungi draw on common sources and serve common ends. It is, again, the chemical web of life, the utter perfection of the natural order.

The partner known as algae can in some lichen species be the dominant form, but in most cases it is not. The algae held in bondage by the fungus on the hemlock where the eagle sat were tamed almost as if they were being cultivated as a slave plant. In nearby ponds and streams grew algae very similar in structure to those forms in the lichen combinations, but not quite the same. While these stream and pond forms can live free and apart from all associations, the forms on the hemlock are as dependent on their fungus hosts as their hosts are on them. They are the Siamese twins of the plant or near-plant world, their union defying the skill of any surgeon's knife. Evolution joined them, and only that same force can separate them, although that does not seem likely to happen.

Even as the lichen spreads from branch to branch and from tree to tree, it is done as a combined effort. The fruiting bodies which fungi normally use to reproduce are largely useless when they come from a lichen union. The lichen reproduces by an action again combined. Tiny clumps form as a few algal cells are entangled in strands of fungal mycelium. The whole carpet of lichen becomes littered by these clumps, and the wind vectors them from place to place. Thus the partners of the lichen association reproduce without benefit of a sexual relationship.

They land and begin to establish new carpets on which new populations of insects will feed, to be fed on, in turn, by other insects and arachnids, this while nitrogen is being drawn from the air and carried to the soil. The relationships of these elements remain largely as mysteries—as each new apparent rule is recognized, new exceptions seem to appear. Yet we can say that this incredible complexity of life on a single standing tree is a universe in miniature. And over all, like a god of that cosmos, the eagle perched and waited until hunger or the urge to play with the wind would launch her free to enter again into a system of her own.

19

Small Creatures and Great Adventures

From
The Forest

So much of what goes on in the places where animals live is hidden, either because the animals and plants are minute or because they are naturally secretive, that the average observer will miss most of what is going on. An untrained hiker is highly unlikely to see a shrew, a rattlesnake, or a magnificent king snake, much less see them interact. The sequence in this next selection would likely be missed entirely.

One thing natural history writers must do is take those organisms that are barely above the level of chemistry themselves and move up their ascending complexity like a ladder until the world of visible yet usually hidden life begins.

All of these things are as much a part of the places where animals live as the grander, more flamboyant species. All start in the same places, down in the chemistry and up in the climate, and all are going toward the same place. (That place, of course, nature writers can't see any more clearly than the most untutored of nature watchers.) It is all tumbling forward, and that forward movement is only really comprehensible when the elements in-

volved are all in their places. Naturalists, for their readers and for
themselves, try to get all of their characters in place.

The tangle of microscopic life that lived within a few millimeters of each of the hemlock's root hairs was a jungle
unto itself, no less a forest than the one in which the hemlock stood. To creatures like the worm, lowly, yet large, all
such life was a mash of organic matter to be ingested with
the soil. Upon such material the worm lived before it fed
the mole that fed the shrew. In quantum leaps the smallest
motes of matter reached the life of the largest and became
part of it. This flow is called the trophic scale, and it is
propelled by the power of the universe itself. The least of
the microcreatures in the root hair jungle was now locked
into the life of the furious shrew, and it was not to stop
there.

A stranger had come to the forest, a visitor from rocky
outcroppings just a little farther up the slope. *Crotalus viri-
dis oreganus*, he was called, the western rattlesnake. Almost
42 inches long, gray and greenish and rough, this specimen
had found no prey for several days. He had slipped over a
rocky ledge as the sun had risen and heated the small natural oven where he was coiled. He had kept coming, heading down the hill. With no fresh scents to entice him or
turn him away, he had moved through the comfortable mid-
day shade, slipping beneath the forest's understory of vine
maple, salmonberry, and devil's club. Now he moved into
the denser shade of the hemlock and Douglas fir stand. It
was moist and quiet, and there were fresh smells, so he
began to cast about. Ceaselessly his tongue moved in and
out, the tiny holders in its twin tips snatching particles of

air, bits of scent, and delivering them to organs in the roof of his mouth. In a way we cannot really understand, he tasted the air, smelled it, too, and came to know its chemistry to an exquisite degree. His awareness was akin to our own in some small ways, but far more acute. For although a dangerous adversary, he was also vulnerable.

Nearby, a larger animal unseen by the rattlesnake moved, and instantly the crawling one reacted. Although deaf to airborne waves of sound, the snake, full out and in contact with the ground, could sense the disturbance. *Cervus elaphus*, the elk passing nearby, was 1,000 pounds heavy, and his vibrations were easy to detect. But the giant moved off, and the snake held his S-shaped striking posture for only another minute before relaxing and going his own way. As always, he had been alert and ready to defend himself, but it suited him better that there had been no showdown, no need to fight and risk his fragile life.

The venom that seeped slowly and constantly into small chambers in the glands at the rear of the snake's upper jaw had evolved from saliva. It had taken millions of years to perfect. Its primary purpose was food getting. Rich in digestive enzymes, like all saliva, it helped the snake stop the flight of fast-moving prey and, simultaneously, began that prey's digestion, even before it died. Only to a minor degree did the venom work on the nervous system of prey animals; its principal target was blood. It was designed to destroy blood tissue and the tissues of the system that carry blood within the animal the snake might choose to eat.

But venom has a secondary use in the relatively few snakes that have it. It can be a means of defense. The rattlesnake had not rattled at the first sign of the elk, but had the disturbance increased, the snake's tail would have begun its frenzied movement. The rattle would have been heard. It would have become the snake's first line of defense. Instinctively a snake would rather hold its venom than waste it on things it cannot eat, so first it rattles to tell beasts like elk and man to move away. It is usually only when that signal fails that the rattlesnake will strike out and inject enough venom to warn the intruder off at a second and higher level of defense. Snakes strike when the whole picture drawn by their combined senses convinces them there is no other choice. It is on such a reflex level that fang use is born.

Still, in one out of four strikes a rattlesnake will not use venom at all. It will rely on the stinging sensation of a *dry* bite to warn the foe of what a rattlesnake can really do. If seriously challenged, though, if injured or teased or badly frightened, the always nervous rattlesnake will deny its natural urge to conserve its fluids and inject a heavy dose of venom, enough, at times, to kill rather than warn its enemies.

The elk had not seen the snake, had not really threatened it, so the encounter had come to nothing. The snake was soon hunting again, searching for prey he could kill with purpose, prey that could help him sustain life. He could eat only what he could swallow whole.

In time the rattlesnake came to the area beneath the hemlock upon which the eagle sat. There was too much growth below the eagle's roost for the bird to see the snake, for indeed she might have considered him a handy, if minor, meal. No snake is a match for an eagle coming in swiftly from on high; none can counter the closing talons striking at 80 miles an hour like hammers equipped with rotating knives.

Undetected, the snake came to an evergreen huckle-berry bush and pulled himself in to wait and watch and see what the moist forest floor would offer. Like the eagle, he was programmed to wait. He hunted equally well in either mode.

Sorex, after she had finished ripping the mole apart and gulping the pieces down, had begun to hunt again. In the time it would take her to locate a fair meal she would be hungry once more. She moved about frenetically under-neath the leaves, snuffling and pushing debris aside. An occasional grub or slug was revealed, and she knew which of these was sweet and which too sour to slay and eat. But they were tiny snacks for the shrew's fierce metabolism.

A shrew's eyesight is not good, for its eyes are extremely small in relation to its skull and suited only to things in close. So the compass of the shrew's world is small, and although it may be frightened by the vibrations of grander forces within its world, it cannot perceive them.

The odor of the shrew is distinctive and well known to rattlesnakes. While not an impressive meal for a large snake, the shrew is warm-blooded and, as protein, entirely acceptable. The snake sensed the disturbance in the leaves nearby and employed special organs in the roof of his mouth, organs known as Jacobson's receptors. Pressing the tips of his tongue into the two cavities there, the snake tasted the shrew at a distance and moved only slightly to be in position to take his meal.

As the meal approached, a second sensing system came into play. Midway between the snake's eyes and external nostril openings lay his heat-sensing pits. That is why his kind are known as pit vipers. Each pit led to two cavities, the larger forward one fed by infrared rays emanating from the shrew. A thin membrane separated the two larger cavities from almost microscopically smaller ones behind. That membrane, rich in nerve endings, measured the heat differential from one side of the snake's head to the other and enabled him to judge his prey's precise position for his bulletlike fang strike. A heat difference of no more than one-fifth of a degree Celsius was enough.

Using both modes—the Jacobson's receptors in his mouth and the infrared radiation entering the cavities in his cheeks—the snake watched the shrew more carefully than the shrew had ever been watched before. When *Sorex* was about 18 inches away, the snake drew back, coil tight upon coil. The spring was charged, and a lifetime of hunting experience would signal the moment of release.

Sorex emerged from behind a small cluster of leaves, barely revealing herself, and a fraction of a second later was rolling over and over in enlarging convulsions. She now contained a substance she had not contained a moment before, a foreign material able to disrupt utterly her entire life system. No power on earth could reverse this explosive chemical shattering of her body. The combined chemistry of the worm, the mole, and the shrew came apart there on the forest floor.

In the instant the shrew had revealed herself, the rattlesnake had tightened one farther inch, moving the top coil only slightly and accentuating the S-shaped curve of the front third of his body. Then he struck. As his head shot forward, his jaws opened so that when they reached the shrew, they would be spread almost 180 degrees. As the upper jaw moved away from the lower, two fangs swung down from their nests on either side and in front of that jaw, and although still almost entirely encased in membranes, they were ready for use. As the snake's perfect strike, guided every millimeter of the way by his heat sensors, reached the shrew, the snake stabbed—he didn't bite, he stabbed, both fangs entering the hot little mammalian body at the same instant. The membranes slid back along the fangs, and the gleaming and highly specialized teeth sank in a full half-inch. Simultaneously the lower jaw of the snake swung shut, gripping the shrew on the point of impalement while bundles of muscles behind the snake's head tightened. The venom moved quickly along ducts leading forward from each gland and entering the fangs on top and in front, above the gum line where the entrance lumina lay. The venom ran down the fangs and exited through ovoid holes again on the fangs' leading edges. It then spread out into tissues of the shrew. She and six unborn young in her would die convulsively within a minute.

All this had taken less than a second. The snake was withdrawing from the strike before the first convulsive shudder ran down the shrew's body. In his withdrawn S position, the snake now held for an additional moment, then began to relax. The shrew rolled over and over nearby until, with eyes hardening, jaws agape, she died. Her young followed in rapid succession, one, then two until all six had ceased to live. The venom had reached them through the minutest of blood vessels connecting them to their mother, had killed them before their mother's own death stopped their lives.

Relaxing still further, the rattlesnake slipped forward to where the shrew lay stiffening. After nosing his kill, the snake opened his jaws a second time and positioned them at the tip of the shrew's nose. He arched his neck. With incurving teeth fixed on the shrew's nose, he began moving in the easy and practiced rhythm of swallowing. In fewer than two minutes the shrew was a small bulge in the slender part of the snake's neck. He slipped away then to a fallen log to digest the shrew and her young before attempting to hunt again. Once more some of the chemistry of the tree and the forest floor had changed ownership. A gentle mist fell as the sun was masked by a moving sheet of clouds. The eagle looked out across the tops of lesser trees and in the distance saw a glimmering. There was the sun farther off, and higher up the slope there was snow.

•　•　•

A nurse tree on which several younger hemlocks grew was slipping slowly into the soil. It was disintegrating. Weather and the roots of other plants, the action of insects as well, had combined to rend it and free its substance. It had become pulpy and soft, not only a nursery for the young of many kinds but also a home for the aged. Containing animal life in every part, it fed everything from microscopic plants and animals to bears that came to pound and hack at it for the beetles and grubs that tunneled through it.

Many of the animals that live in such a sheltered nest are called thigmotactic. That word describes an animal tropism, a negative or positive incentive, something that attracts or repels a living thing. Light is the basis of two tropisms: the phototropic are attracted by light and move toward it, while the photophobic (the earthworm is an example) move away. Thigmotaxis is such a tropism. It is the desire to be touched on all sides at the same time. Thigmotactic creatures live under bark and under rocks. Often we disturb them and see them frantically scurry to rebury themselves. We assume it is light or temperature or some other force we inflict that worries them. Most often it is not that, but rather a force we have denied them: the force of being touched all over and always at the same time. For that is the force of their comfort and safety. They evolved needing it, for its need meant a higher survival rate.

Many animals are not truly thigmotactic, but at times do seek such security. Under the nurse log, where the roots of four young hemlocks had taken their hold, lived or sometimes just rested many animal forms, all more or less addicted to the tropism of touch.

On the day the rattlesnake came to pay his uncommon visit to the world beneath the hemlock, a creature of great beauty was using the nurse log to comfort himself and hold himself safe—*Lampropeltis zonata*, the exquisite, glossy California mountain king snake. Almost 40 inches long, he lay coiled, quietly resting while faint signals told him it would soon be time to hunt. The rat he had taken four days before had been immature, small, and was all but used up. While not ravening in his hunger, the king snake was close to being stimulated into hunting again. In the meantime, he held his coils, tucked back in against an overhang of undigested log, safely touched and waiting.

The king snake of the California mountains is one of the most beautiful of all the world's snakes. Shiny and smooth, it has jet black as its first color, over the forward two-thirds of its head. Then comes a narrow band of creamy white, then one of black jet again. A broad band of brilliant red follows, and so it goes the length of the king snake's tubular body: narrow black, narrow cream, narrow black, and broad red. Some people liken it to the coral snake of the South, but no coral snake lives in California, and so no dangerous confusion is possible. The king snake, unlike the coral snake, bears no venom. It is harmless except to the small animals on which it preys.

To the scientist the king snake is in part ophiophagous, an eater of other snakes as well as of mammals and birds. Its generic name, *Lampropeltis*, means "shiny skin." As a group (there are only eight species), the king snakes belong only to the New World and range from Canada to Equador. None is more beautiful than *L. zonata*, the king snake of California's mountains.

Although *L. zonata* may be found at sea level (but never in the desert), they are more common at higher altitudes. They may, in fact, be found as high as 8,000 feet—jewels, chains of glistening, muscular light and color, hunting, resting, and hunting again. The king snake hunts during the daylit hours unless the weather is very hot. Then it will shift its pattern and hunt at night.

On this day he lay coiled not far from where the rattlesnake had taken *Sorex*. The disturbance had not been lost to the king snake. An old and experienced warrior, he knew the vibrations that came to him along the forest floor, and he knew the scents, for his forked tongue, too, flicked in and out to taste the air and learn its offerings and its threats.

The musklike odor of the rattlesnake was easy to detect, and the almost hungry king snake was triggered. The scent had shifted the balance, and now he was ready to eat—to hunt and *then* to eat. The taste of the rattlesnake on the air was no less than would be the smell of a steak on the grill or a cake in the oven to a human nose. The king snake uncoiled, stretched out flat, and started out tight in where the flaking undercurve of the log met the debris-strewn floor. Guided by the intensity of smell and by the vibrations transmitted through the ground, he moved toward an intersecting path. Extended straight out in order to keep close to the log and in the gray light of the falling mist, the king snake gleamed like a rope of gems more than 3 feet long. In fact almost 2 inches shorter than the rattlesnake, he would still cast the challenge when they met. He was immune to rattlesnake venom and could thus afford the chance. He might be driven off by a frenzied defense of some sort, but he could not be killed by an animal whose only real weapon was an injectable chemical.

The king snake remained hidden during the period of the rattlesnake's hunt, kill, and feeding. Only when the rattlesnake moved away to coil beneath a bush did the gleaming hunter begin to stalk his meal.

As the king snake had located the rattler through his powers of chemoreception, the rattler now used the same means to detect the mounting danger. Within minutes of the rattlesnake's coming to rest, the two hunters faced each other with eyes that could not blink and lips that could neither smile nor express fear or pain.

The rattlesnake feared the king snake by instinct and therefore did not assume the normal defensive striking coil or S shape. Instead, he pushed his aftercoils against the stem of the bush under which he had crawled, thereby to gain leverage for a backward twining retreat, and pressed his head and the front third of his body against the ground. His nervous pitch had peaked in the first few seconds of awareness, and now his tail buzzed angrily. The king snake moved forward slightly and then did a strange thing. Coiling loosely, he began to vibrate his tail as well. Resting as it did among some leaves, the tip set up a buzz not unlike that of the rattlesnake's horny tail. The two snakes answered each other's warning *whrrrring*, although neither could hear. It was a strange reflexive mute play in the forest's understory gloom. The king snake buzzed his false rattle sound for only a moment, then pressed on, for the rattlesnake was pulling away, writhing backward over a fallen branch. The king snake followed until the rattler was pressed back tight against a large log, unable to get over it without raising his head off the ground. And this the rattler was unwilling to do. It would be the head the king snake would try to grasp to begin swallowing.

The king snake came within a foot of the rattlesnake before the venomous one made his first truly defensive move. Instead of striking out, as he would have done against any other animal that threatened him, the rattler looped the middle third of his heavy, roughly scaled body and raised it high over his head. Then he slithered forward, and instantly the heavy coil of his tense and threatened body struck the ground near the king snake's head. The king snake pulled back as the rattlesnake again turned the center of his body into a looped club and raised it to swing down once more.

Over and over the rattlesnake's body thumped to the ground, once striking the king snake a glancing blow to the head when he failed to withdraw in time. Each time the king snake backed off. In what he had inherited within the substance of his egg was the certain knowledge that his patience was greater, that he could outwait the rattlesnake. His prey would tire. The king snake expended almost no energy at all. The rattler was exhausting himself with his lashing attack. His only hope was that the king snake might become intimidated and move away. This the glistening warrior was not about to do. He had met too many rattle-snakes in his life, and only when he had been very small had the danger of a battered skull driven him away. As a large snake he had come to accept his supremacy and take his hysterical prey with cold-eyed efficiency.

Within minutes the rattler had indeed worn himself down, and the lashing coil fell less frequently to earth before the king snake's nose, and with less conviction. Having been struck only once, and that a minor blow, the king snake was undeterred. He moved back and forth, in and out, drawing, yet avoiding the thumping punishment of the rattlesnake's nonchemical defense.

Nearby, on the other side of the nurse log, there sat another stranger to this part of the forest. Having pushed aside the log's rotting wood in search of grubs, *Euarctos americanus*, the American black bear, was sitting back on his haunches and munching on the treasure hoard he had uncovered. He listened as the rattlesnake's coil struck the ground again and again. Unable to see the action nearby, from time to time he pointed his muzzle straight to the sky and wrinkled his nose. He snuffled and sniffed, for like all bears, he believed only his nose. Deciding that there would be time enough to investigate, he went back to his grubs, making sucking sounds over the joy of their texture and their slightly sharp taste. His long claws flaked away layers of tree growth, now pulped free of many structural chemicals, in order to expose more of the unformed insects. He would finish here first, before looking into whatever was taking place no more than two dozen feet away.

Hundreds of feet above all this, the eagle sat waiting for the signal that would tell her to kick free again and seek her own meal. Again she looked up the slope to where snow was melting, sending rivulets down toward the relatively level floor of the coniferous stand where the hemlock grew.

The rattlesnake was tired now. Two dozen times he had raised his looped body and cast it toward the king snake only to strike the hard ground. As the moving target of the king snake had drawn him out, the debris underneath the rattlesnake had become more unsettled. His only fixed point of contact was his chin on the forest floor. He slithered back and forth but did not relinquish that one hold on safety. Touch, contact, is always important to a snake. At a moment of extreme peril it is more important than ever.

The king snake's instincts triumphed. The resolve of the rattlesnake finally dissipated. No matter where he pushed and coiled the enemy was always there, blocking him. A rattlesnake deprived of the use of his venom, and instinctively aware of that fact, can be worn down easily. The king snake knew this with his own instincts, taught by time and the million rattlesnakes taken by his ancestors, and when the moment was right, he rushed. Like a gleaming arrow, he struck and had the rattlesnake's usually deadly head in his front teeth. He coiled around the venomous one and constricted. The rattlesnake could not fill his lungs, and soon his heart stopped beating. Working back and forth, hinged in front and at the center, the king snake's bottom jaw worked over the broader span of his lance-headed prey. Although in death he still lashed and coiled in one ineffective move after another, the rattlesnake was soon on his way down toward the pool of powerful digestive juices within the king snake, juices that would break down not only the snake but also the shrew the snake had eaten, and the mole, and the worm, and the microscopic mash that had fed the worm.

The bear snuffled up over the log, tired of beetles and their grubs and curious about the disturbance nearby. The eagle tipped forward, for now she sensed the bear. But there was no meal there, nothing the eagle could challenge and kill. *Kya*, her falsetto rang out, *kya, kya, kya*.

❧ 20

A Case of Energetics

From
The Endless Migrations

One of the most amazing of natural phenomena in nature is the imperative and ability of animals to change their places—to migrate toward safer places and better survival opportunities in response to changing weather patterns. I have made a few selections from The Endless Migrations *to show how widespread this pattern is—this mass movement of life-forms from the very simple to the most complex away from bad weather toward good conditions including available food and water. The cost to the migrators in energy is enormous. But migration serves another great role. It quickly weeds out the weak and the inferior young born every year, leaving only the best examples of a species to return the following year and reproduce.*

During the summer months tiny crystals of protein life called zooplankton clustered and pulsed near the surface of the waters surrounding the continent. So diverse in form that hundreds of thousands are yet to be named, sparkling in a rainbow of electric color, these minute creatures, some neither plant nor truly animal, fed nonstop on the phytoplankton, or equally minute plantlike life, that flourished in the same rich broth of the sea. Copepods and the larval form of many fish constituted the bulk of the zooplankton, but the entire mass spread and rose and fell in response to light intensity like a shimmering blanket suspended in the ocean. Vertical currents brought nutrient-rich water up from the floor of the sea, and the phytoplankton fed on these chemicals and reproduced as rapidly as the zooplankton diminished their store. All this was in equilibrium, and a passing whale straining tons of the blanket away at each long swallow did not make so much as a furrow in the immense sea meadows undulating in every direction.

But, then, the changing season was sensed here in the oceans as well. The phytoplankton's rate of increase began to fall rapidly as the sunlit hours diminished in number and each day the zooplankton began sinking deeper into the sea. At some level, yards or even miles below, the zooplankton would survive the winter, not feeding but in the timeless patience of the living drift, waiting for the warmer season to return and for the plant meadows under the sea wind to explode into incalculable numbers. The up-and-down movement of the zooplankton was no less a migration than the thousands of miles to be traveled by the whales and the birds and the butterflies, because migration is movement. It is an essential flow with a single function, the enhancement of survival through constant adaptation. Animals and even plants move for that reason alone—to survive, and for each species survival demands different energetics, different skills, different fitness tests.

Bats are among the least of mammals. Only the shrews

are smaller than the smallest of the bats, yet even at half an ounce the little brown bat faces an autumnal migration. All those that summer across New England move westward to one of only four known caves. For some it is a journey of two hundred miles. The caves—in Connecticut, Pennsylvania, New York, and Vermont—gather their hoard in the fall as temperatures diminish and then discharge them again in the spring. The repeated journey and the long winter fast do little harm, for some have life-spans exceeding a quarter of a century.

The guano bats, *Tadarida*, have a different pattern. In the summer they range across all of southern North America. For six weeks in late winter, the sexes mingle and mating occurs. Later, from mid-May to the end of July, the sexes form separate colonies, generally staying in different caves. They remain apart until the young are born and have shown an ability to survive, for at birth small bats face danger from below. If they fail in their first attempts to fly, voracious beetles scurrying in the darkness through the detritus on the cave floor below wait for the weak or the imperfect to fall so they can set upon them and tear them apart. Those infants that survive must face another early trial. Unlike the little brown bats of New England that migrate two hundred miles or less, the guano bats must fly from southern North America to central Argentina and Chile and then return north the following spring. The distances between summer roosts and wintering sites measure many thousands of miles.

The barren-ground caribou, *Rangifer*, the deer of the far north, has established individual herds across the top of North America. The Western Arctic Herd, the Central Arctic Herd, the herds known as McKinley, Delta, and

Nelchina are all totally within Alaska. Two other herds, the Fortymile and Porcupine, disregard international boundaries in the way of migrating animals and range in both Alaska and Canada's Yukon. Then there are herds that are wholly within Canadian territory—the Bluenose, Bathurst, Beverly, Kaminuriak, Wager Bay, Melville Peninsula, and Baffin.

The movements of the once great caribou herds, which carried them across highways and near villages and towns, placed them on a collision course with man. In the middle 1960s two of the Alaskan herds, the Fortymile and the Nelchina, numbered fifty thousand and sixty thousand animals each. Ten years later there were only remnants left: five thousand animals in the Fortymile herd and perhaps ten thousand in the Nelchina.

The role of the wolf in caribou survival during migration has puzzled man for decades and brought friends to the brink of battle. It is generally true that such predators as wolves do not control the populations of their prey. Quite the opposite is true. The population of prey animals dictates how many predators will be born, will grow, and will live to hunt. Since the Arctic wolves, with their incredible music of the wild night and the storm, live for much of their lives almost exclusively on caribou, the balance between those truths is vital to the survival of both species.

There are those who say—and this has become fashionable—that the wolves take only the weak, the sick, and the already doomed from the fringes of the moving herds. Surely to some degree that is true, and herds are healthier for the loss of the inferior individuals in their midst. These could live to breed and create more low-grade animals, eventually whittling the herd down through loss of quality.

Herds of animals, like steel, must be tempered, and for the caribou the wolf is both the flame and the anvil.

But there are storms in the Arctic, storms of such intensity that they cannot even be imagined farther south; there are swollen rivers that send streams of caribou thundering across the landscape to their death in the jaws of the wind; there are inaccessible food and even cold when fat reserves have fallen too low. There are scientists who say the wind and the snow and the river crossings are enough of a culling force and the wolves serve little function if any. But, then, there are those who issue and those who get licenses to hunt wolves from the air and who sell their pelts for sporting wear. The battles for and against the wolf run on like a leitmotif of the north, but still they and the caribou survive and press on in cycles more ancient to the Arctic than man, cycles as old as the wolf. The fact that man found both wolf and caribou together when he pushed toward the north is proof, one would think, that the two species belong together, one feeding the other, one serving that upon which it feeds in ways we understand imperfectly.

The caribou is an animal of elastic strengths. It is resilient and adapts well to wolf and man, even though man has added to his arsenal at an exponential rate. The wolves, however, stood still; their technology of death did not improve. Their tactics were ancient, their needs unchanging, they had no commerce, no new reasons for taking more animals in new ways. The wolf killed, strenuously, to live while man superimposed the supposed legitimacy of sport and after that the vague claim of integrity and salvation. Still the caribou traveled past the firing lines, thundered past settlements, even used roads and railbeds where they existed on the soggy tundra. Their numbers continued to fall, but without the power of oversight and with no under-standing or concern for their own population dynamics, the caribou pushed on. They were locked into ways as un-yielding as that of the salmon, the monarch butterfly, and the gray whale. Although opportunistic and adaptable when cleared roads offered a path in the right direction, still the caribou adopted their direction first and held to that rigidly no matter what they encountered.

Even in the farthest reaches of the north there is a time of joy. Summer heat and the curtains of hungry insects are gone for the year and the harshness is still weeks away. The days are warm enough for every animal to move with ease and pleasure, yet cool evenings signal the time ahead and frost glitters across the landscape to greet each morning's sun like enormous ice sheets shattered and strewn. At tra-ditional places, usually in sheltered valleys where trails have been trodden for thousands of years, cranberries and blueberries and other autumn crops ripen. Late salmon runs fill the streams. Ptarmigan move toward higher ground in anticipation of the coming snow and the sky fills with waterfowl seeking their winter ranges far to the south. Swans cluster in surprising thousands, with mallard, gold-eneye, and mergansers all among them. Widgeon, old squaw, ducks of sweet water and of the sea, moving south-ward and facing thousands of miles of flight before coming to rest for yet another winter, are all part of one vast cur-rent in the sky. The Arctic is like a huge engine of life, and each fall it pulses and pours forth living product and lends it to distant and warmer lands, though under contract for a single season only.

All manner of birds facing monumental flights were now layered with fat, and so indeed were the caribou that began to move, thickly furred and armed against the harshness from which they could not flee on wings. They would face the ice storms, become half buried in snow-driven winds, yet stand fast to live and reproduce again the following spring. The birds had fat for flight, which they would burn at the rapid pace of avian metabolism, while the slower mammals would build the corresponding fat to help them hold in place, slowly absorbing what they needed while using the outstanding balance for insulation.

As autumn colors blazed across the tundra, the largest of the caribou found it difficult to regulate body temperatures through the layers of back fat that held in the heat. They breathed rapidly, flicked their ears against the last, lingering black flies, and ate some more. More fat was needed; every ounce was an investment against the poor death of a harsh winter and an unprepared animal. The winter that was coming was a force without limit. Nothing could stop it. The caribou could only prepare and wait and eat beyond the demands of immediate need. The drive to engorge was the hunger of life, the need to live and to move the pattern of caribou life forward for yet another generation.

There is a herd mindlessness to animals in mass movement. A single caribou or a dozen might turn to stare at a pair of wolves and worry about their stalking movements. A few caribou might flee at the mere sight of hunters. But when the hoofed creatures form into vast floods, when they coalesce into massed thousands, the mind of the single animal is overwhelmed, absorbed by the sheer magnitude of the many. A herd of thousands of animals will ignore the wolves. The mass presses on, profoundly content that it can spare its losses. A single animal cannot spare some part of itself, a single cow cannot spare her calf, but a large herd can. It is the quandary of the mass mind, the herdiness that dulls the senses of the individual. Perhaps, and we speculate here, there are leaders at the front of the movement who worry about the wolves on the horizon ahead, but behind them come the thousands that are made so secure by their fellow creatures pressing in through touch and scent and sound that they are mindless, pounding toward winter and the drive to survive it.

The thundering caribou may veer slightly or even occasionally turn if their leaders turn, but generally, in their mindless push toward the coming season, the herd ignores wolves, hunters, all else, and listens only to some primal force. That force, the coming of migration, controls the mass mind of the herd. And when the antlered herds and the flights of birds and the plankton swarms do not appear, the predators in turn die in the agony of hunger, cannibalism, and disease.

The caribou followed an ancient pattern. At the end of July the big summer herds had splintered, moved apart. The animals had gone off singly or in small herds, bachelors, cows, and the young alone at last, each to find others of their kind and to feed leisurely together in the time of Arctic plenty. But as the air began to chill, first at night and with each successive day later into morning, the animals began coming together again. Two bulls, solitary for much of the summer, began feeding near each other, then were joined by two more. Their antlers were now full grown, and it remained only to strip away the velvet that concealed their new hardness and their pointed tips. Bulls gathered near every tree, every bush, in the all but treeless land and rubbed against the branches. Bubbles of blood showed through the velvet, but the soft cover fell away in tatters, and the vegetable sap and berry juice and the blood from the tiny vessels of the now fallen velvet slowly darkened the antlers as they were tempered for battle. Farther south other deer were following the same pattern, and nearby in overlapping range the giant of all deer, the moose, were also embarked on the same task.

Where two caribou had grazed there were ten, then a hundred. After wandering in circles for several days they began to move with purpose. Those on high ground flowed down toward the river banks, where the vegetation would last longer and survive the night frosts. In the willow thickets they encountered the moose and moved away, leaving the solitary giants the peace and space they demanded. Rivers appeared and were crossed, some torrents running hard enough to claim lives. Cows and immature animals were now flowing in the same general direction as the bulls, but the bulls still kept apart. The bands, not yet a large massed herd, were often in sight of one another, and both began to encounter wolves, creatures that had learned down the millennia where the caravan would invariably crest a certain river bank, where they would compress their growing masses through ancient cuts in the land.

And now the bulls began shoving one another, engaged in what appeared to be good-natured games. Their future lay in combat for the possession of the cows. So the bulls moved and mock-fought and moved some more. Immature animals and cows maintained their parallel tracks.

Each night the frost froze harder, each morning it was slower to disperse. Then one night a blizzard struck. The animals and the land were snow-covered by the time the late-rising sun appeared. The colors of the tundra, explosive only a few days or a week before, were faded and gone. Now the landscape was white. As the caribou thundered on, their numbers growing by the hour, flights of birds crossed high overhead, pushing south before the next blizzard struck. The calls of the geese and swans and lesser birds echoed against the ground and whistled through the wind like one musical theme that is working its way through another. It was a fabric of sound that came and went, whistled and hummed, until the symphony of descending winter encompassed all.

The immense pattern of lakes and ponds that dotted the tundra were now frozen, but there were few with ice thick enough to sustain the pounding of a caribou herd. The animals crashed through the splintering ice, and frozen water squirted up around them like diamond dust and silver rain. A few animals foundered, but most pushed on. It was only a matter of a few more days before temperatures became consistently low enough to turn the water into a pavement as hard as anything that man could create. The lakes and ponds and rivers would turn from umber and earth to silver, blue, and white, and the herds would use them as avenues of convenience.

Hunters formed their ranks along the migratory path, fired into the herds, and moved about later collecting their kill. Those wounded animals that managed to pull ahead of the hunters and stay with the herd would begin to fall back in a day or two. Wolves would soon pull them down, but those that fell before the wolf packs struck would be found by bears, grizzlies of the barren ground that rolled on the carcasses to absorb their smell, then snuggled their great snouts into the entrails to pull them apart at leisure. No part of a caribou would be wasted. Magpies, crows, ravens, and foxes would move in on the fallen animals as well and pull away flesh that had come to them so effortlessly.

By the middle of October a restlessness began to move through the congregation of bulls. They grunted rhythmically, their necks now swollen, their sharpened antlers set for battle. Younger bulls learned quickly that they were outclassed, and only the larger, fuller, and more brutally aroused animals remained in position to battle for the cows. Skidding and slipping, tearing up turf and snow with antlers and hooves, the bulls fought across the increasingly broken ground. The triumphant bulls moved off to join the now receptive cows. In the meantime the caribou continued to move, their paths intercepting, their numbers growing, and their purpose ever more clear as they moved to where the snow cover might be thin, the lichens and grasses beneath more readily grazed. It was now a time to survive. Mating occurred, young were promised for another year, but now the winter must be endured—and this, then, was the real reason for the migration. As the mating eased off, the bulls wandered away, joined each other, sought deep, quiet cuts where there was any vegetation at all. It was cold every hour now; every minute of the day and night it would become colder still. The herd was about to face the ultimate test. The best would emerge in the spring and disperse back toward the areas from which the thundering mass had come. A circle was half drawn.

The migration of the caribou covered hundreds of miles, making it quite probably the longest overland migration by a terrestrial animal now known. They moved from an area where willow branches had been stripped to one where grasses and lichens could still be anticipated under a covering of snow. They followed that primordial route because their instincts, honed over thousands of generations, had pushed them to it. But to know to move, to know to seek food, to know the time is right is not necessarily to know the way. There lies a mystery not yet solved.

There is good reason to believe that magnetism is a factor, perhaps the most profound factor, but we do not know for certain that such is the case. The earth's position had changed, and the animals had changed in relation to the sun, so there could be visual clues. There are prevailing winds, possibly half-remembered land shapes and forms, altitudes in sequence, and, in the higher latitudes, cosmic radiation far more intense than in areas near the equator. Incoming radiation strikes the earth's magnetic shield and slides around the planet like quicksilver beads on a giant slope, coming to earth near the poles. Intensity of radiation may offer clues, but we cannot say for certain how the caribou find their way. They do find it, though, surely, unerringly, in predictable wave upon wave, to complete that first half of the circle. It is the first half because it is during that movement that mating, the future, is assured.

🌸 21

The Fabric
of Migration

From
The Endless Migrations

Once certain basics have been worked into a natural history story, things like chemistry and weather mechanics and instincts, the writer owes the reader a tale. Naturalists may have clarion horns to sound, important messages to transmit while there is still something to write about in our games of show-and-tell, but writers are still storytellers. Writers who lose sight of the storytelling part of the profession lose sight of themselves and probably lose not only their sense of humor but begin to take themselves far too seriously. Readers can sense that and may quickly opt to go somewhere else, to different authors and perhaps to different subjects. The writers then lose not only a reader and a walking advertisement for their work but they may lose a convert. Not everybody who reads a book on nature is a confirmed nature buff. Some people are exploring, checking the subject out to see what all the fuss is about. They are our recruits. Old-time nature buffs are certainly where we focus much of our attention but there is a touch of incest in that. We want fresh fields, new faces, opportunities to hook new energy by recruitment. For that task we need not only animals properly situated in their places but animals in full display of their power, their might, and in action. We must always be storytellers.

The finest of the threads in the fabric of migration is pulled and tugged through the loom by the least of feathered animals, the one-eighth-ounce hummingbirds. On this vast planet the smallest of birds, they are found only in the New World. The pugnacious, rapacious little sprites gather wherever there is good feeding. They prefer red flowers, but they will take the nectar they can find and insects as well. Their metabolism is atomic and they must feed much of the time. Each a jewel, each a small electric explosion of color and iridescence, the birds must feed a furnace and will gain in return incomparable powers of flight.

The male ruby-throated hummingbird, easternmost member of his family in North America, was three inches long and weighed but a fraction of an ounce, no more than that. Metallic green above, he had a glowing ruby throat that marked both his sex and his maturity. Remarkably, he beat his wings seventy times every second. It appeared as if he were suspended in a veil of shimmering gauze. His body was distinct, but his wings were only a blur. He could hover seemingly endlessly, he could fly sideways, backward, and straight up and straight down without noticeably changing his body attitude. His needlelike bill darted in and out of flowers, sucking in nectar and occasionally the smallest of insects that hid deep in the crevasse of a flower's heart.

It was autumn; it was growing cooler each day and the hours of sunlight were shortening. Like billions of other

222

birds around the globe, the ruby-throated hummingbird had received the signal "Store fat." Other birds might store it by the ounce or even, in a few cases, the pound, but the ruby-throated hummingbird lays it on in weights equal to those of postage stamps. Still, it was that minute store of fat that would enable the sprite to accomplish an epic journey. The bird—actually a tropical bird that had managed to work its way north rather than a northern species that wintered in strange lands to the south—would seek Mexico, where virtually all North American hummingbirds, even the widespread ruby-throateds from as far away as Alaska, go when winter comes. Deep into Mexico they go, to areas least touched by the weather. They consume energy in vast quantities relative to their size and must have food, and food for them means budding flowers, nectar, nature's sugar.

From immediately after sunup until just before dawn the little mote sought out fall flowers and sucked in nectar. He consumed as well a slightly higher weight in insects than during the humid, rich summer months, when nectar was less difficult to find, when, in fact, it was everywhere in endless quantities. He fed and he fed, and the minute layer of fat built and built as it must if he was to live. The monumental journey lay ahead, to be revealed by his genes.

One morning the ruby-throated awoke with first light, moved off his perch deep in the heart of a bush, and spun in place. Like a bullet he was gone. He reached fifty miles an hour almost immediately and vanished from the field at the edge of which he had spent his huddled night. The field was behind him, then the town, and then the next, and with his wings a blur he was across the state and heading southwest. He would reach the coast, the northern rim of the

Gulf of Mexico, and then face five hundred miles of open water. At no point in that five hundred miles could he set down for so much as an instant.

On the way to the Gulf's margin he fed. He would drop into a field, hover before a flower, drive his needle-sharp bill forward into its innermost parts, and suck, then be gone. He seldom alighted during the day, but he did settle deep in thick vegetation at night. With each first light he would be gone again—a small memory that left no imprint. He may have weighed no more than a goose's feather, but everything the migrating goose had packed into it was tucked into this incredibly small unit of life as well. Mexico, an ancient memory, beckoned in so powerful a voice, his whole body had become attuned to response. He would take the states of the United States one at a time, he would take the vast Gulf of Mexico, and arrive home at last, having come before the wind, before the storm, before the weather that would surely have ended his life if it had ever reached him for so much as an hour. He was a tropical bird en route home. He might breed in the north—indeed, he had been hatched on Long Island—but he was a tropical jewel on loan and was being called home, where his colors would blaze until it was safe to venture north on loan again.

He finally crossed the boundary and was at sea. There was a different taste to the air and a new feel to the wind. There were also new problems in navigation, for all features disappeared the moment he crossed the line between land and water. Rivers carrying silt down from the continent to the north colored the water a uniform clay. Only far beyond would it be blue again. Small, choppy gusts of wind eddied and brought salty mist and then just salt into the air through which the hummingbird darted southward.

But he was still bulletlike, fired and aimed toward his southern genetic home.

The little bird's course was toward the southwest so that he would come ashore on Quintana Roo Territory, on the easternmost outreach of the Yucatán Peninsula, beyond an ancient barrier reef. He would drop briefly on the Isle of Cozumel, feed, then push due west for less than an hour, and finally come ashore where the Mayan Indians so long ago built the cliff-top city of Tulum.

Then, over Mexican territory at last, he would turn south, feeding at the edge of jungles and over marshes that followed one after another. In the forest itself, not far from the sea, giant pyramids and temples stood draped in green, almost totally overgrown with vines and brush, and many of these species bore flowers. The energy lost over the featureless Gulf was quickly replaced in swamps and jungles where ancient civilizations once flourished.

The hummingbird's ability to hover depended on a network of muscles that constituted nearly a quarter of the little animal's total weight. In relationship to the whole animal, though, the hummingbird's wing muscles were huge. His wings acted as propellers when he fed from a flower, for he had to hover, fly without moving. He hung from his wings much the way a helicopter does from its blades. The stroke was downward and forward, then upward and backward, with the bird's body inclined forward at a steep angle. The breath driven downward by the wings at maximum effort was about the same as the draft created by a languidly waving palm leaf on a hot, still, and humid day. The hummingbird's ability to fly sideways as well as backward, the nature of its figure-eight wing stroke, make it resemble a large flying insect more than a tiny bird.

The little ruby-throated bird had a kind of computer

built into its flight mechanism which told it how much energy to expend for each maneuver and how much fat to burn. In lateral movements the wingbeat varied from forty-two to seventy or more beats every second; hovering under different circumstances also told it how many beats to permit and how powerful each beat would have to be.

When the hummingbird stopped to feed, it did so because of its own sense of how much fat it had consumed and how much more was critical for its continued flight. It was judging fuel reserves as critically as it was energy expenditure, managing a fine-tuned network of nerves and their genetically coded demands.

The journey, even at fifty miles an hour during the sunlit hours, required all of the bird's strength and its navigational skills as well. It did not use ground signals, but it had the sense of its diagonal flight across the path of the sun, and that was tuned perfectly to the time of the year. Other hummingbirds, which had wintered and mated and built nests of cobwebs in other parts of the species' range, flew at different angles to the sun and were prepared as well for what was required of them in direction adjustments. As to how birds driven off course by winds on either journey—to the north in the spring, to the south in the fall—could ever compensate for that angle across the sun's track is a mystery far from solved. These wondrous creatures inherit not only information about geography and celestial maps but their ability to adapt the data in an infinite number of ways.

The speed of its flight created problems for the bird, and for these other compensations were necessary. His wings moved so fast in creating their blur that friction was intense, and that friction meant energy loss. Food had to be consumed and fat stored for the sole purpose of overcoming friction, and it was a considerable demand. The wing oscillations, those parts of the cycles that were more insectlike than birdlike, created eddies, which, again, had to be accounted for. Nothing the bird did or could do was without its price, and the price was fat, fat ready to be burned and then replaced. The hummingbird, as it sought the Mexico of invisible memory, was a balancing mechanism. It was a flashing jewel, a speck of colored light, a sunspot in a blurred gauze veil of flight-sustaining motion. More than that, however, it was a balancer of calories, a mechanical wonder that wandered over salt water and land of many altitudes, dipping in to feed, then vanishing like a fairy light, remote yet common, seeable yet unexplainable, one of the miracles of nature.

As the hummingbirds dispersed into swamps and meadows of the Yucatán new threats appeared that were unknown at the northern ends of their cycles. Coming in to feed from a flower whose widespread petals promised nectar, the rubythroat was met with the open jaws of a lunging arboreal snake poised near the flower, awaiting its meal. It knew the hummingbirds were coming. It missed, and the ruby-throated darted away so fast that the snake wasn't sure which direction it had taken. But the snake drew back and prepared to wait again, for the bright red-brown rufous hummingbird with its flame-red throat was coming too. The purple-throated Lucifer hummingbird would follow in days, as would the black-chinned, with the blue-violet band on its throat. There would be a few calliope hummers, with purple-red rays against a white field on their throats, and others too.

The birds would come singly, but come they would, to

spend the winter darting from flower to flower in the bewildering world of the tropics. Some would push farther into Central America, but all would leave the storms and uncertainties of the United States and Canada behind. Where the orchids clustered on tiger-striped stalks ten feet high the hummers would avoid snakes and even flights of aggressive bees and at night, as they rested, blankets of marauding ants. Safety from weather and starvation came at a high price, and the hummingbirds, as they gathered from the lands to the north, had only their speed and ability in maneuvering to compensate for each new threat, plus memory triggered by new hours of sunlight, new temperature and humidity, and perhaps the nature of available food. The hummingbird near the Guatemala border is a different bird from the one outside of Boston or San Francisco. It has to be to survive.

The hummingbird, no less than the monarch butterfly or the tern or the goose, has to read its environment, because all actions of all migrating animals are in some way timed and tuned to the world that immediately surrounds them. There are beacons, there must be forces that say abandon this and seek the other for now it is time. Photoperiodism—the ability to respond to the photoperiod, or hours of light received daily—is known to be important for many species. Not all, but many. Without question, a shortening day spells one kind of behavior, a lengthening day demands another. Hours of sunlight not only mean signals, they create conditions that can be survived, conditions that can kill, and those best avoided although sublethal. Sunlight means warmth, both the periods when it is available and the angle at which it is received.

.   .   .

And as the terns of the high north settled into their high southern retreat, the hummingbirds that had in fact abandoned easier places pushed deeper into the jungles and forests of Central and South America. Migrating species are seen at each end of their journey, in each of their two principal habitats, as half animals only. The ruby-throated hummingbird that had spent part of its year on Long Island and was now sparkling from vine to vine in pursuit of nectar from totally different species of plants was one animal in the north and almost another in the south. In the south it would not experience the urge to mate. It would not ravage cobwebs, ignoring spiders in their threat displays in order to build the softest of nests. It would select from a totally different botanical world and be alert to different dangers. The twining tree snakes and massive spiders did not exist on Long Island, but in the tropical forest the hummer had to be alert to them every time it selected a promising flower whose nectar it needed to support its incredible demands for near-instant energy. That part didn't change, that was constant. The tiny bird still burned fuel like a fire storm.

As the hummer had come south, it was as if there were a television set automatically changing channels. The picture changed from day to day, often from hour to hour, as did the opportunities and the challenges. One moment there would be flowers, shrubs, trees, places to rest and feed so that the journey could continue. Then a coast would pass below and there would be nothing of any use to the bird at all except a following wind. For hours, open water, dangerous eddies of wind, and still no food. The hummer began dipping into its fat reserves, and just when exhaustion threatened, just as the fuel was about to run out, another coast would appear, and the tiny mote of sun and feathers would drop low, seek a blossom, and feed on the wing. At times it would land on a bush and rest, but often it simply fed and rose and shot out of sight like a dart from a bow, heading farther south yet.

The route the hummer took had been evolved genetically so the coast would always appear before it was too late. Hummers that tried other routes, that impressed upon their genes course memories that did not include the coast and the flowers, in time perished. They did not return to the north, they did not breed and did not pass their memories along, for they were deadly ones. Their latitudes had been the same, but the coasts and the flowers did not comply.

The hummingbird from Long Island spotted a bell-like flower hanging from a vine in a softly shaded glade of endless tones of green. The flower was red and glistening with dampness. The mite flitted first to one side of the prize and then to another, and then dropped several inches so that it could approach from below.

Pushing upward, it tilted on the axis of its wings and thrust its long spike of a beak up into the heart of the flower. It did not see the angry bee until it was too late to avoid it. The bee came out of the flower buzzing its rage and stung the bird once on the side of the head with the spike at the rear end of its abdomen. The hummingbird sickened immediately, for, although it was twice the size of the bee, it had been assaulted chemically. Wobbling, struggling to stay on a balanced course, it dropped away to a brushy mid-trunk stem on a hardwood tropical forest tree. It caught its perch as it was about to slip by and held on, fighting to steady itself. The venom the bee had injected had reached the bird's nervous system and the hummer's vision began to blur, and then, even as it still hung on, it went blind. The bird tipped forward and back fighting the sickness in the only way it could, by remaining on its perch. If it let go it would fall to the cushion below. The fall would not kill it, but the ants and beetles that scurried about down there would soon tear it apart.

Blind as it was, the hummer could not see the snake as it slipped across from a bush nearby. The snake did not allow its weight to rest on the same shoot that supported the hummer, for it too had instincts and knew in that way that its weight would tilt the branch and signal its approach. It struck from its mid-air position and took the blind hummer off its perch, barely disturbing the leaves that surrounded but did not adequately protect it. The hummer would not breed again on Long Island. After its epic journey, after braving the seas and coastal winds, after crossing through scores of gradations of habitat types, it had been killed by a bee and a snake, species found only at one end of the journey, species belonging to only one half the little bird's life. It had never encountered either species before, but there are no excuses when death is the reward for failure.

❦ 22

The Silent Manatee

From
The Endless Migrations

Some animals are so remote from most people's experiences and very often are so rare that they cry out for inclusion. They provide the writer with an opportunity to recreate not only a place but an unusual creature that gives that place a special dimension.

Florida has many strange and wonderful native inhabitants and a startling number of introduced ones that like it there, thank you very much! But one of the strangest Floridians of all is a wonderful harmless creature whose nature is so benign that it is being crowded off the edge of our planet. The manatee, in place or in motion between places, is part of the wonder of one of our greatest natural wonderlands.

Far, far to the south of any range ever known by a polar bear, another marine animal of great bulk follows a course dictated by its need for food and warm water. From the southernmost stretch of the South Carolina coast, down the east coast of the Floridian peninsula, around its tip and up the west coast, then west to Texas, when the weather and the water are warm enough, the manatee is to be found. A slow, calm, quiet animal, it must surface to breathe, which it does with a particularly pleasing and relaxed, sighing sound. A vegetarian, it harms nothing, competes with no one, but is often threatened by chance and carelessness and the changes wrought by the advent of time. Long ago its ancestors, perhaps related to the animals from which our cows arose, gave up the land for warm seas. Since accomplishing that adjustment, the biology of the manatee has been reluctant to make any other.

The adult manatee sensed the first signs of discomfort. It had spent the summer in a shallow lagoon fed by freshwater springs inland from Florida's northeast coast. Years before, a ship channel had been gouged out of the river bottom, and it was this channel the manatee had followed in from the coast. It had had no trouble adjusting for salty, then brackish, then sweet fresh water. It maneuvered its massive thirteen-foot bulk easily and slowly and nibbled at water weeds by pulling the water hyacinth down from below. It hung in the water and two or three times every five to seven minutes pushed its nostrils up through the barrier into the air above.

But now there was discomfort. A cold front had brought the winter temperature down to 60 degrees, and that was the threshold of peril. Fourteen degrees lower and the water would become a deathtrap. Despite its bulk the manatee had little tolerance. It survived in a world of narrow margins. At 46 degrees above zero the manatee would die. Aware of this in some primitive way, it began moving out of the shallow lagoon, sighing softly to bring its calf to its side. They began moving toward the coast. Once they reached salt water they would turn south, slip around the tip of the Floridian peninsula, and seek warmer waters in the Gulf. That was their only means of survival—slow-motion flight before cold eddies and currents. That is how they had been programmed by their time in the sea as refugees from competition on land.

Hour by hour the taste of salt in the water increased and the thugging, crumping sounds of boats became more frequent. The pair stayed close to the river bank and nosed through the shallow water, through the water weeds, as each mechanical monster violated their quiet world, their submerged world of peace. Intrusive sound is vulgar and harsh, and the manatees recoiled from it almost as if they knew that to be true.

On one occasion an immense oceangoing tug churned up the channel as it pushed a huge barge of bulk cargo ahead of it. The blades of the tug's screw were large enough to chop a manatee in half, but the animals kept clear and pushed into the salty front that lay ahead, seeking both warmth and peace.

Then, on the fourth day, the ocean opened up before them. They worked their way out through the river's estuary, staying to shallow water to avoid the thunderous noise of the shipping and lighter power craft that pulsed in from all directions. They turned south, and then, suddenly, the water began to warm rapidly. A few hundred yards farther the temperature rose more than 20 degrees and they floated into an area where over a hundred of their kind hung in contentment, with their single-lobed flukes lower than their heads. They had arrived at an outspill from a power plant, and hot water continuously poured into the seas, creating a serendipitous manatee refuge. And there were no powerboats, no shipping, just warmth free of intrusion.

It was an artificial lagoon ideal for manatee, ideal but for one factor. Beyond the heated bay the temperature of the seawater continued to drop. Cold water was held back by the constant outflow of the power plant, but the temperature beyond rapidly dropped to some 50 degrees as a northern front began moving a cold current ahead of it. Finally the water beyond the lagoon reached 45 degrees. It was now an invisible barrier as strong as steel fencing holding the manatees to their sanctuary. The cold water could not penetrate the ten-acre space; the sluice flow was too steady, too heavy for that. There was another factor, however. Within two weeks the plant was scheduled to be shut down for long-term repairs and, in part, conversion. The manatees could all die. They could be inundated when the sluices stopped and the cold water moved in on them like a giant club—this unless the temperature of the water beyond the outspill area rose sharply before their sanctuary dissolved.

Far out to sea, undulating toward the north, where it would eventually swing east, was the Gulf Stream. It was warm enough to keep most of the British Isles and even Iceland temperate, more temperate than New York City even though that city is on the same latitude as Madrid. The Gulf Stream is a capricious current and is at times forced seaward by cold masses moving down the coast of the Carolinas, Georgia, and Florida. Then, again, it will swing in toward land, and that is what happened a week after the manatee cow and her calf found the haven of the power station discharge area. The cold wave passed and lost itself in the Caribbean, and the Gulf Stream arced inward as the coastal waters began to warm. By the time the power station began closing down, as the discharge slowed to a trickle, the water beyond the artificial bay was warm enough for the manatee to move south along the coast toward the string of islands known as the Keys. The barrier was down, and the first to move were those closest to the mouth of the bay. They slipped out and turned south automatically. Soon others followed, and within two days no more than a dozen manatees remained in the inlet. In time they would leave as well.

The cow and her calf pushed south toward Miami Beach, Key Biscayne, and the other islands beyond. Almost at random they wandered into shallow rivers and canals, hung close to shore, and fed on the succulent underwater vegetation. They fed mostly at night, but their soft sighing at the surface could be heard at any hour. If a particularly noisy powerboat came screaming through a canal close to where they hung suspended they dropped toward the muddy bottom. They could stay submerged for fifteen minutes without distress, but before sixteen minutes had passed they had to breathe again. Normally they breathed at five-minute intervals, but that was when there was no apparent danger nearby. They lived their lives in slow motion and simply retired from anything existing at any other pace.

The two animals stayed close together, usually touching, as they drifted under bridges and passed private docks and the gaping mouths of marinas, from which their foes the powerboats spewed forth. They brushed against buoys and markers, making them bob in the water and nod almost politely as old friends passed by below unseen. Gulls and ducks and other water riders looked down, only mildly interested, as the hulks moved beneath them. Wading birds—egrets, herons, and their kin—some migrating visitors from much farther north, stopped stalking frogs and crayfish and other prey long enough to acknowledge the passage with a call of *kraaaaankkk* or *cawaw* or other sounds more shrill.

Mother and daughter pushed on toward warmth and food, keeping to themselves and the water's edge. Canals suited them well, for they were dredged and had sharp corners below rather than rapidly ascending shores like a natural lake or stream. Very often they went for two or even three days without feeding, for the same dredging that gave them a safe bottom contour often destroyed plant life. But there was food enough, and the temperature held, and at intervals the two travelers emerged into salty water and pushed along the coast again. They drifted back and forth between rivers, canals, and small lakes and the sea. When they reached the first Keys south of the peninsula's tip, they passed seaward of them and, south of Tavernier, moved through the cut toward the west and emerged into the Florida Bay part of the Gulf of Mexico. Barely clear of the Keys, they began moving north, passing just offshore, west of the islands whose eastern shores they had passed a week before. And again they reached peninsular Florida, and their journey northward continued. They chose at random from canals, ponds, small rivers, and the ocean off the beach.

Once they had passed Sanibel Island they moved close to shore, passing between Pine and Lacosta islands. For several days they fed in a tangled river system and then came back to the Gulf again south of Englewood, near Gasparilla Island. Every day was marked with hazard, as their trip south had been, for boats shrieked by, swung around mangrove islands, and bore down on them at speeds surpassing anything they could attain. Their only defense was to sink, and they did that at the first pulse of an intruding craft. By sinking and by holding as close as possible to every shore, every bank, they avoided the wrenching impact that had already killed more than a dozen of the manatees that had taken refuge at the foot of the east coast power plant weeks before. The calf observed her mother and learned of hazard by experiencing it with her. Now she did not wait for her parent to react. She had learned the great secret of manatee survival, and at the first warning sound she sank on her own, although her mother was never far behind.

At one point they passed close to a group of six blue sharks, not one less than ten feet long. But the sharks had just fed on offal dumped from a fleet of fishing craft and were too lazy to attack. The manatees moved as close to shore as the bottom grade would allow and there was no engagement. Had there been, it would have been fatal for at least the calf and perhaps the mother as well. Following the near encounter with the sharks the manatees moved into a river and hung there for several days. Small crocodile, rare in Florida but still to be found in some areas, ignored them, as did the more frequently encountered alligators. Most were far too small to attempt a prize of their size. And then they reached the northern edge of the Gulf, where both Florida and the coast of North America swing to the west. Their course too was toward the west, Alabama and Louisiana, which lay far ahead. The water was warm, and the epic journey of the somnambulant manatees would succeed another time.

As each mile unfolds before a migrating animal the potential for new hazards is laid bare. Not only are the worlds through which they pass ever-changing sequences but the fact of their passage, of their passing from one uncertain environment to another, compounds the chances for surprise and intrusion. The manatee's world, particularly, since it is so inherently serene, is open to the shock of invasion and collision. Temperature, the mechanics of the human world, other animals, natural obstacles, all are potential enemies.

One last dangerous encounter awaited the mother manatee and her calf before they reached the peace of their warm wintering grounds. A shrimp trawler laced out its nets across the opening between two small islands. Beyond was a harbor almost completely enclosed by land, and in certain weather conditions clouds of shrimp moved in from offshore and pulsed through the sea like a gray-pink tide. A drag here could gather up tons of the small animals and dump them, squirming, onto the deck for quick sorting by machine and freezing once the boat brought them to the water-edge processing plant only a score of miles away.

The manatees could hear the shrimper and feel the vibrations in the water as the now alarmed mass of shrimp ran before the net. But the manatees could not understand what was happening or from which direction the real hazard came. The net was moving toward them and bore with it the potential to entangle them and pull them under and hold them there longer than their lungs could endure. A netted manatee drowns as readily as any land animal would. Manatees do not eat shrimp, but they moved toward them and then through the cloud, for the greater menace seemed to be coming at them from another direction. They swam directly toward the net, and then, as it so often will, the sea offered a surprise ending. Just as the net was about to engulf the manatees in an incredible mass of confused and tumbling shrimp, the bottom of the net hooked onto the ribs of a broken boat that had died there in a hurricane years earlier. For fifty feet of its two-mile length the net was pulled below the surface and held there, with twelve feet of clear free water above it, before its bottom rocked the skeletal boat and broke free its floats to carry it back into position. In those few minutes a river of shrimp flowed over the top edge of the net into the free water beyond, and the manatee mother and calf flowed with them, flopped across, and dipped down toward the bottom and pumped their flukes until the net and its capacity to kill were well behind them. Without ever comprehending the danger they had faced, they escaped it and moved on toward the west.

When cold fronts stopped moving down into the northern areas of Florida with their accompanying effects on water temperature, the manatees would sense that they had to retrace their journey. They would move east, then south, then east and north again. Theirs too was a rhythm, and their return to the east and north was an act of unconscious faith. All migratory animals must act on faith, for they die if what they have learned to expect by genetic memory fails them. They cannot call ahead, there is no all-clear signal, just timing and faith. It is a faith on which the universe is built.

❧ 23

The End of the Pleistocene

From
Mara Simba the African Lion

When a foreign or exotic setting is to be the backdrop for an entire story, the writer can properly assume that most of his or her readers will not have been there. The place is almost surely stocked with strange vegetation and strange animals in balance with each other. Although there will be other elements that can wait to be introduced later on, on a bit-by-bit basis, the central cast of characters and the way they are tied to each other and to their place is the first order of business. I think it is best to jump right in. And so we have here an introduction to the Mara Maasai.

On a personal note, my wife Jill and I first went to the Mara with Joy Adamson (Born Free) in early 1971. I had decided to do a book about lions and had decided from what I had read that that was a particularly good place. It is in Kenya, it gets the Serengeti migration, it has plenty of lions and is reasonably accessible from Nairobi. I suspected I would have to make more than one trip. I was right. I went there twenty-two times over a period of thirteen years before daring to undertake passing my *impressions along to others. It was a far cry from trying to write about the Kodiak bear on Kodiak Island without having been there at all!*

The Mara Maasai is a place, all right, with a presence, an ambiance, a texture that comes close to defying description. I like to think that it can and has been described. The first part of that process follows.

The people call it *E-Mururuai*, "the place of the sacrifice, the holy place." No one is quite sure why it came to be called that, but it is a name from *opam oitie*, "very long ago." It must not be changed, for the powers that made the place from out of nothingness would be disturbed and perhaps angry. Since it is far older than human memory, it is sacrosanct. It is a shallow cut between two low hills, in the Narok District of southern Kenya, just south of a place called *Ang'ata Naado*, "the long plain." It is in the reserve called *Maasai Mara*, and the indigenous people are the Maasai. They are the almost legendary pastoral nomads of the central portion of the Great Rift Valley. The Rift is, except for the vast ocean basins, the largest geological feature on earth. It extends on a diagonal, southwest to northeast, from Southwest Africa to at least the Middle East or all the way to the center of the Soviet Union, depending on which expert's standards you accept.

As the enormous Rift Valley slashes across Tanzania and up into Kenya it smooths out into the rolling Serengeti Plains. The plains are known as Serengeti only in Tanzania; where they climb north into glorious uplands the word Serengeti vanishes, a victim of European politics. Before World War I a line was drawn just south of the upland sweep to keep colonial Germans and British apart. The Serengeti Uplands became the Maasai Mara in Kenya, English territory.

The Maasai tribesmen have never understood European politics, although they have always understood the Europeans themselves much better than most tourists and not a few explorers and missionaries believed. So, the Maasai have drifted back and forth with their flocks and herds across what, to them, is a wholly meaningless boundary. It has gone on for centuries, at least since the 1600s when pastoralists speaking a language known as Kalenjin, or perhaps it was Maa back then, moved south from far drier lands in the north. They moved out onto the plains where rivers ran toward cuts in distant hills and the grass was as high as the armpits of a tall warrior. The Maasai and their herds and flocks drifted with the wet and dry seasons, millions of head of game beside them, and the lions fed on both domestic and wild animals. Probably then known as *il-Maa*, they moved onto the stage where the last act of the Ice Age was being played out. In time, probably because people who visit Africa need tales to tell, the Maasai became fabled as warriors, stock raiders, and the inveterate enemies of peaceful endeavor. Most of that is fiction. The Maasai are proud, stubborn, and perhaps even arrogant, but they always accepted a peaceful alternative if it was available to them. They would rather talk than fight.

The Maasai Mara, or just Mara as it is generally known today, is not quite like any other place on earth. Thousands of feet below the rim of the Great Rift, deep in a valley fifty miles wide, the still seismically active grasslands provide fodder for almost uncountable numbers and varieties of animals. Nowhere are life and death played out on a grander scale or against a more magnificent backdrop. Everything about the Mara is vast, grand, awe-inspiring. It is little wonder that the Maasai came, stayed, and say to this day, *Meeta entoki marisiore*, "There is not a thing which is equal to it." With a tenacity matching that of the wild herds that preceded them, the Maasai have hung on and defied time. In fact, time has passed the herds, the Maasai, and the lions by. In a sense, the Mara is a time capsule.

Throughout the vast expanses of the Serengeti to the south and the Mara itself, the predominant influence is grass—or, rather, grasses, for their species are many. The grasses are like an enormous musical instrument with the wind the musician. Grass represents the perfect integration of climatic factors and soil condition, and the many Mara species are quick to settle, colonize, and dominate any available area, sandwiched in as they are between deserts to the north and forests to the south. These tropical grasslands constitute one of the richest wildlife habitats this planet has ever known. It is a place bursting with life, a belt of life just south of another belt, the equator.

Technically, the Mara is a savannah, which means that the mean distance between the trunks of the trees that grow there, however many there may be in any zone, will always be greater than the diameter of the green canopy overhead.

Grasses of many kinds feed humans and animals all over this planet, and the Mara has more than its share of the basic kinds. Among them are annuals, plants whose seeds may lie dormant for months or even years waiting for ideal conditions so that they can germinate quickly and cover the hills and fill the valleys with nutrition. The important substances are in the seeds and the animals know to crop them there and to ignore the relatively less useful stems nearer the ground. In the perennial grasses growth superficially resembles the annuals, but these grasses do not blossom as quickly, and their nutritional value is in their stems or even underground in the rhizomes or corms by which the plants spread and reproduce. Their seeds are incidental food to the animals that harvest them. In bad times, when there is drought, the perennials do not wait it out with unsprouted seeds. They transfer their nutritional value to their underground parts where the weather matters less. They have the power to do this, to switch the location of their chemical reserves to another part of the plant.

The actual species count of the Mara grasses, like that of the birds and the mammals overhead and aboveground, is almost without limit. In the uplands at least twenty species of bluestem grasses grow, tall and tufted perennials. The needle grasses number twenty-four species, their seeds ending in sharp awns. Signal grass, Rhodes grass, the protein-rich star grass, and love grass are an incredible mosaic, like a bewildering green and gold buffet upon which millions of animals of thousands of kinds can feed. Hood grass, a robust tufted perennial and the grass from which the local people make thatch to cover their cottages, feeds millions of animals. There are sixty forms of guinea grass; there are thirty species of napier grass, sometimes called Kikuyu grass, thirty species of *Setaria*, and forty species of dropseed. All are part of the mosaic, all filling niches, all combining the goodness of soil, water, and the sun for everything from bacteria to elephants to feed on. That is the grassland of the Mara: trillions of individual plants packaging the energy of the cosmos and making it accessible to billions of animals.

Mixed in among the grasses are the legumes and the herbs: twenty species of *Acalypha*, ten of the flowering *Astripomoea*, twenty of *Cleome*, at least twenty of *Desmodium*, twenty of the *Dolichos* legumes, one hundred fifty of *Indigofera*. Herbs and legumes among the grasses constitute a table set and waiting to which the incomparable herds of migrating animals come in an eternal cycle as linked to the rains as are the grasses themselves.

The grasses, legumes, and herbs are at their best when rains fill the void between land and sky. The herds of hoofed animals can sense falling water from a hundred miles away and come to it as if drawn by an inexorable force. There is no winter, no summer in the Mara, hard by the equator as it is, but there are two seasons at opposite ends of a spectrum—wet and dry. When it is dry in one place the animals stream to another that can offer what they need. From August until late October, virtually every year, the herds are in the north, in the Mara. As many as one and three-quarters million of the wildebeest, white-bearded gnu, move up from the lower Serengeti, across the border from Tanzania. Like the Maasai they ignore boundary lines that originated in London and Berlin. With them are as many as half a million zebras, common or Burchell's zebras with broad white and black stripes that go right around their bellies. European and American tourists come and see a white animal with black stripes that the Maasai know to be a black animal with white stripes. There are reedbuck, defassa waterbuck, eland, topi, kongoni (or Coke's hartebeest), and hyrax. There are uncountable impala, Thomson's gazelle, Grant's gazelle, and Peters's, too. The "Tommies" stand around stamping their tiny feet, wagging their tails incessantly, and looking for all the world like millions of windup toys—and very attractive ones at that. There are warthog, bushbuck, dik-dik, baboon, and ostrich.

In wooded areas, back between hidden hills, there are giraffes, the Maasai variety with leafy-edged brown-red blotches, thousands of the three-thousand-pound giants wandering back wherever trees grow, for they feed high, not on grass. There are rhinoceros in the Mara, but fewer now because of the heavy traffic in their horns and other parts. The Moslems of India believe that a ring made of rhinoceros skin worn on the right hand will cure hemorrhoids. In India they believe that a man dipping his parts in rhinoceros urine will perform his masculine role with greater pride and zest. Vendors outside houses of prostitution provide beakers of urine to entering clients. For these absurd reasons mighty rhinoceroses die at a rate that has brought them to the edge of extinction. In India the "street value" of the horn from a single rhinoceros is about ninety thousand dollars, for it is said to cure virtually all ills.

The rhinos will soon be gone from the Mara because Somali poachers have worked their way down through the entire length of Kenya to hunt in the Mara, rob it of its rhinoceros, and shoot at game wardens who get in their way. It is not a land at peace, as the Maasai would prefer it. Greed has come to the land of grasses, herbs, and legumes. It is not a greed that has arisen there like everything else, from the soil. It is a foreign greed feeding the insatiable madness of foreign markets. The Maasai hunt only predators, like the lion, to protect their herds, and are not a significant party to the senseless destruction that has invaded their land. Poachers and anti-poachers use aircraft and hunt each other with automatic weapons. The Maasai and their herds stand aside, apart, but are called warlike.

There are elephants in the Mara, too, often large herds of them, and the signs of their destructive ways are everywhere. A small herd can convert a forest to a savannah in less than a quarter of a century. At times only a few elephants, the *il-tomia* of the Maasai, will be seen, but then, suddenly, there will be forty, the bulk of the herd, if such it is, appearing from nowhere. Elephants for all their enormous size are like apparitions. Ghostlike, they drift into a valley, eat, quietly uproot trees, pull up grass with their trunks, and then are gone like the wind that sighs through the grass. It is certain that they have gone toward the rain, but their movements are secret and often unseen. That anything so large can come and go like a wraith in eerie silence is in itself awe-inspiring.

It is its taste for the nutrition-rich grasses that eventually does the elephant in. Each elephant grows seven sets of teeth in its lifetime. It pulls grass up by the roots, and the grains of sand that come up with the plants grind down the elephant's teeth and make the multiple new sets essential to an animal that may live between sixty and seventy years. When the last set of teeth is so badly worn that the elephant can no longer process tough hillside grasses, the animal moves toward water where the plants will be softer, making the giant less dependent on its ability to grind and chew. Since elephants digest only half the food they consume, vast quantities of food must be eaten every day. The day comes when the elephant crosses a line: It no longer has the strength to process the amount of green matter it must have, and it starves to death or simply deteriorates. Most dead elephants are found near marshes and other areas where pulpy waterside vegetation, much of it giant grass, can grow. Even in old age the elephant is not attacked by other animals. It is one of the very few animals in the world that dies of old age. There are no cats left on earth that can attack an animal that is between eight and eleven feet high at the shoulders.

Cape buffalo inhabit the Mara by the thousands. Peaceful when in herds, and these herds often contain hundreds of animals, the great buffs are unaccountably cranky and extremely dangerous when alone, particularly the bulls. That is when the Maasai must fear them, when the lone bull buff steps out from behind a tree. He is likely to charge, anxious to kill, made mean in part, perhaps, because of the lions that stalk the herd constantly paying no mind to how big the buff are. Many lions die trying to kill big buff, but many buff are killed by lions. Each can feed a whole pride of lions. When lions and the buff meet it is a fearsome collision. The gods of the grass must choose between titans.

In the deeper-cut rivers of the Mara there are hippopotamus, great lumbering gray beasts that are far more aggressive than the maligned rhino who want only to be left in peace. The hippos claim plots along the riverbank and territories with easy access to the breeding females who gather in an area known as the creche. The best spots are held by the biggest and most aggressive bulls. The constant "yawning" motions portrayed in so many tourist photographs and seen on postcards are not yawns at all. They are threat gestures from bulls to all other bulls, and even to men, on the riverbank. They say, simply, "Stay away." A boat drifting through hippo territory is likely to be bitten in half if it crosses into a mature bull's riverine claim when cows are near. The bull hippopotamus does not distinguish between species. Only its claim matters, and its strength. That strength is primeval, and so is the dull wit that controls it. Its purpose is reproduction and so serves the species.

Two cats of substantial size besides the lion prey on the herds of the Mara. These are the leopard and the cheetah. Leopards, or *ol-owuaru oti*, can weigh two hundred pounds, but usually they weigh considerably less. They ambush game from trees; they stalk it relentlessly and carry their prizes up into acacia trees where other cats cannot come, for the leopard is the Mara's only true climbing cat of any size. Lions may rest on lower limbs to escape biting flies, but that is another thing.

The cheetahs of the Mara are plentiful and are the swiftest mammals of all. With little stamina but with the incredible power to move from a standing-stalking pose to a run of almost seventy miles an hour in a few seconds, they take the smaller gazelles. They must drag their prize behind a fallen log to hide from marauders or eat it where they have made their kill and take their chances. Life is dangerous for their cubs.

There are hyena in force, secretive python, flotillas of crocodiles, packs of Cape hunting dogs, jackal, bat-eared fox, and cats called serval and caracal, the latter related to the bobcat and lynx. Small African wildcats hunt at night. There are eagles of several kinds in great numbers and squadrons of vultures and marabou storks to clean up after the other meat eaters. At night, genet and civet (relatives of the mongoose), weasel, and banded mongoose themselves hunt and stalk and kill a variety of prey. Cobra, puff adder, and many other snakes both venomous and harmless hunt their small insect, bird, and rodent prey. Underground, endless tunnel complexes and nurseries mark the termitoria, a single nest often containing several million animals. Snakes find their way into the surface protrusions of these so-called anthills (in fact, termites are not at all related to ants and are much older and more primitive in structure, although not in social organization) and slither out into the grass to be preyed upon by aberrant eagles on stilts, secretary birds, who hunt the snakes and kill the most venomous of them by stomping on them and pecking them like jackhammers. The birds are agile enough to dodge even a spitting cobra.

The lion has been called the king of the jungle as well as king of the beasts. The first error in those romanticized names is readily apparent by picturing the grassy Mara, which is ideal lion country. The lion is an animal of the savannah, an animal at the top of the food scale as a hunter. Never a jungle cat, rarely even a forest cat, the lion lives in the open like the cheetah and trims the top off a food chain that starts down among the grass seed plumes and stems. As for *king*, that, too, is a fallacy. The lion will attack a rhinoceros only under the most extreme conditions and then only the youngest, one whose mother has been killed. A lion could not kill an adult rhinoceros or an adult hippopotamus under any circumstances and will give way to either.

As for the real king, surely it is the elephant, for everything gives way before this mountain of strength and cunning. The lions of the Mara fit in, feed well, work hard much of the year, and have easy pickings only when the great herds have moved in from the south. The lion is a prince, certainly, but not a king, not in the long grass, not when wild dogs and hyenas can tear a tired old lion apart, not when a slight misjudgment or moment of bad timing can turn a buffalo hunt into a lion slaughter. The lion even has reason to fear pythons and crocodiles and venomous snakes as well. A lion does well to survive in the long grass of the Mara, but a king he never was.

24

Death and Dying,
a Way of Life

From
Mara Simba the African Lion

Why, some people ask, must animals in stories die? Because the lovely painting created (again and again, somewhere between sixty and ninety times) by Edward Hicks, The Peaceable Kingdom, *is a fantasy. We kill to eat and so do a great many other animals. Very, very few animals, in fact, escape one of the two roles. Elephants are neither predators nor prey except under the most unusual circumstances where a large cat finds a calf unattended (almost unheard of) but most other animals are one or the other, or both. Certainly lions are prey. The old cats are regularly killed and eaten by hyenas and Cape hunting dog packs, and the mortality rate among cubs is around 50 percent.*

Death, then, is part of the life of any animal, and if an element in the animal's environment is the agent of that death then it belongs in any story truly told.

The lioness walked away from the cubs, yawned again, grunted four times with a thunder from within that rattled the very rocks of the natal hideout, and then voided. She instinctively did it far enough away from her cubs so as not to provide a beacon for other predators even though she was about to move her nest. She wandered down toward the opening where the small marsh began, stood looking out over the plains, and grunted five more times. In the distance the other females of her pride would hear her and then understand, although she spoke to the world at large.

Behind her her cubs crawled over each other, rolling often and only beginning to sense each other's presence. It was their first chance to explore the world beyond the womb and it would be the pattern of their lives. Their curiosity would make them teachable. Their ability to learn the social graces of the world's only true wild-cat society would be the key to their survival, although the percentages against each were tremendous. It would require a near miracle—rather, a series of miracles—for any one of them to survive the fact that each was as much usable protein as protein user.

The number of different kinds of animals that had slept that night in the cut between the hills at *E-Mururuai* was almost astronomical if insects, arachnids, and the cryptic species of the soil are taken into account. Many were tiny, hiding species that had to be touched on all sides simultaneously to feel secure. Scientists call such animals thigmotactic, and they are seldom seen although they, too, live in a world of fierce competition. But not all animals that hide and want to be in contact with their surroundings in the most intimate way are tiny. At least one is a magnificently handsome giant of startling proportions. The Maasai have long known the glistening beauty of *en-tara*, which scientists have named *Python sebae*. It is the African rock python, a monster that can grow to thirty-two feet in length and be almost as big around as a grown man. The python is a sly hunter of immense power and unblinking eyes.

Nowhere in the world are giant snakes worshipped as they are in Africa. People not far from where the Maasai wander speak of the god Danhg-bi when they speak of this snake. It has been revered since times so ancient no one can reckon their place in the evolution of culture. Captive specimens are kept in temples of grass and wood and consulted, propitiated, and tenderly cared for as long as they live. Virgins serve them; priests and keepers with secret powers speak with them about matters of eternity.

The python of *E-Mururuai* was a wild, free hunter who depended on his glossy light-brown scales marked with blendings of darker brown and random speckling to hide him until he was close enough to strike his virtually irresistible blow. He could dart forward from his own giant coils at nearly six miles an hour. Although tales of his kind butting prey and knocking it down with the force of the strike are almost certainly nonsense, the python is equipped with immensely powerful jaws, and once it clamps down on prey with its inward-curving teeth, the victim seldom escapes. Prey is chosen with care, for like all snakes the python is fragile, subject to infections and parasites; to live it has to avoid wounds that could be invaded and become septic. A snake, too, even a giant snake, has muscles like steel bands, but they are supported by a relatively light skeleton. A python is a giant that can be torn as well as broken.

During the night the python of *E-Mururuai* had taken a hyrax higher up on the side of the southern slope. It had been a simple kill, but it had also been a very small meal for an animal so large. With the coming of light more protein was needed. The giant slipped off the limb onto a ledge, easily spanning the open space between, moved smoothly along the rim rock, and entered a deep stand of grass on an incline that led down toward the mouth of the cut and the inevitably profitable marsh beyond. His senses told him that there were no animals in the cut itself that were worth hunting. He would move to the bottom and then out toward the marsh and select from what was there beyond. It was a path he had taken often before. He seldom failed when he began hunting in earnest. His forked tongue sampled the molecules of air and inserted its take into an organ in the roof of his mouth. He tasted the world as he moved his great bulk forward in the slow, inexorable rhythm of a death messenger.

The lioness had made her decision. On the far side of the marsh there was a small stand of trees, acacia and ficus mostly. There was shade there and a good view of the rangeland below and its constantly changing stock of prey animals. Of the nearly seventy species of acacia in Africa, seven occur in the Mara, and three of these were in the area the lioness chose from a distance. From there she could spot game she would be able to hunt by herself until she was ready to rejoin her pride with her young. From there, too, she could spot and avoid or drive off tough young males that might be too lazy to hunt for themselves and might try to cannibalize her cubs. The intervening marsh was soggy, but that would not deter her from crossing it, taking each of her cubs in turn to the new nest site. Although her life now seemed to revolve around her cubs and their needs she knew instinctively that her own survival had to come first. She had to eat and maintain her condition or both she and her cubs would die. She could produce more cubs, but her cubs could not find another mother.

She returned to the cubs and rolled the smallest of the four over onto its belly, prodding it with her nose. It squealed softly and grunted as she took its entire head into her mouth. It instinctively drew its hind legs up against its belly. She trotted off with it across the soggy ground. It flopped and jiggled but did not struggle. She found a shallow depression filled with red oat grass, grass ungrazed, deep enough to hide her four secrets and, she hoped, keep them from harm. In their first few hours of life they had become her total preoccupation. She dropped the infant into the grass-lined shallow depression and trotted back to her three remaining cubs. The next two responded as the first had, by drawing up their hind legs and not struggling. Three cubs were soon piled together in their new nest, all grunting and squealing and struggling to right themselves and find their mother. They were all hungry and were sniffling and prodding trying to find the nipple none had yet seen but from which each had fed and which had become for each the focus of life itself. Nothing else really mattered, for when they were not nursing they were sleeping.

The largest of the four cubs, the one that weighed five pounds at birth, was the last still in its place of birth when the python slipped out of the deep grass twenty feet away. Instantly sensing movement ahead, the python increased his speed, tasting the air as he went until he was only a few feet away from the cub. At that point he lunged, for he had already identified his target. He dragged the cub toward him as he moved his own coils forward and lashed them around the barely struggling infant. In seconds his first coils closed over the cub and began drawing in like an enormous knot. No bones were broken; the cub was not crushed, for that is not the way the constricting snakes kill. As the little cat exhaled, the coils tightened, leaving less room inside for the lungs to expand. In less than two minutes the heart of the already unconscious cub stopped beating.

Before the python could position himself to swallow the cub headfirst, the lioness arrived to retrieve the last of her young. The cub was gone. It took her less than thirty seconds of casting around the area to locate the python. Her snarl came like an exploding boiler. Her lips were back and it was as if fire came from her yellow eyes. The python sensed his dangerous enemy as soon as she saw him and uncoiled, abandoning the cub he had just killed. Two killers were now stalking each other, one no less dangerous than the other. To be certain of victory each needed the advantage of surprise. A python, unseen, can drop on even an adult lion and perhaps kill it. A cat coming upon a python that has just fed and is semi-torpid might surprise it and win. But these two killers were alert, awake, facing each other. Although neither could feel hate as an emotion, they could manifest rage. They were dangerously aroused adversaries. The hormones that had seen the lioness through her pregnancy and birth turned her into a dragon at the disturbance of her nest site, while the true dragon, the mighty python, had been disturbed at his meal.

The snake lashed out, his jaws gaping, but he was far slower than the enraged lioness. On his first thrust the python was met with the smashing impact of a great paw. Before the paw was withdrawn claws had curved out of their sheaths, making the first raking cuts that would spell death for the snake eventually, even if the cat withdrew. But the cat moved forward and hammered the snake again and again with a rage born in the mists of ancient times. The snake writhed, tried to strike out at the cat, threw his coils into loops and piles. But his spine was broken not far down from his neck. His beautiful shiny scales were matted with his own fluids, were ripped and soiled with blood-soaked earth, and still the lioness hammered. Finally, as the snake's movements slowed the cat moved close enough to take the snake's head in her mouth. There was a crunching sound. The primitive nervous system would keep the snake writhing there, twitching and jerking, for hours. Hyenas, jackals, and vultures would eat the snake and scatter his light, fragile bones. In less than twenty-four hours there would be nothing left but disturbed grass and some blood mixed with soil.

The lioness walked over to her largest cub, sniffed and prodded him, and then picked him up by the head as she had her other young. This cub did not draw up his hind legs but hung limp, his legs dragging as his mother moved out across the soggy marsh. Halfway across she abandoned her dead infant and moved off without him, breaking into a trot to reach her remaining young sooner. She made a low, moaning sound as she moved. Later in the day a hyena would find the cub and carry it back to her underground den to feed her own cubs. The first cub had died without ever seeing light, but three more remained. Who can say what, if anything, the lioness felt? We have neither the insight nor the language. She would have killed the python if she had encountered it under any circumstances. If she did not have cubs waiting unattended, she might have eaten some of the snake, although she would have left most of it for scavengers to use.

The python weighed two hundred thirty-six pounds and was over twenty-five feet long. The African valley that had seen its birth began to absorb it back into its chemical cycle just as it did the largest of the four new cubs. It was as if they had never been. As it is with every living thing in the great savannah they had been temporary holding systems for a complex chemical matrix. But at the end, and sometimes very near the beginning, it must all break back down. The python and the lion were agents of that redistribution.

25

The River Pride

From
Mara Simba the African Lion

Very intelligent animals like the great cats select places for at least temporary use according to a hierarchy of needs and uses. Water must be available and prey must not be too far off so that the young cats can be taken on a hunt or food brought back to those not yet ready to travel even short distances. In the case of lions there must be a reasonable distance between the place under consideration and those occupied by other prides. Space is critical, and it is only by respecting each other's space that deadly dangerous animals like lions can keep real fighting to an almost irreducible minimum. Real fighting on any kind of sustained basis would mean the end of the lion species. There has to be a mediating influence and that influence is space.

Place, then, meets biological needs and social imperatives. It is far from the only thing animals seek in life but it is one of the most important. An animal out of place is in deep trouble. For many reasons it may not survive. It is hard enough to survive when all the needs of a place are met.

A river flowed across the savannah running from the northeast to the southwest. It caught some of the wash from the eastern escarpment of the Great Rift Valley, but much of its force came from deep underground. In a series of holes and caves at the foot of the towering cliffs that marked the edge of the Valley the water bubbled up, formed several marshes, and came back together again in ever more discrete channels. Great lakes were strung out like jewels not many miles beyond the boundaries of the Valley: Elementata, Naivasha, Nakuru, Bagorio, Baringo, and finally Lake Victoria. Billions of gallons of water were gathered into these natural impoundments and countless billions more flowed underground, under deserts and savannahs, under mountains, and even under the lakes themselves. In some places the water reached the surface, as it did at the head of the Narokajia River.

The course of the river was meandering at first, but as it moved south it flowed between increasingly higher banks and gathered force as well. By the time it reached the general area where the young males hunted and sought their destiny it was nearly two hundred yards across, although for much of the year most of that expanse was sand. At times the river became several small streams flowing through a cut the parent river had made and then it would widen again in other seasons when rainstorms lashed across the savannah.

Riverine islands and sandbars were resting places for crocodiles and a number of stalking birds. Saddlebill storks; egrets; East African crowned cranes; goliath herons, the largest herons in the world; and many lesser species were a few of the many. Tall trees—those that had survived the assaults of elephants—lined the banks, and heavily worn trails led down at an angle showing where the elephants came daily to drink and bathe when they were in the area. Crocodiles ranging from a foot or even less to almost fourteen feet in length could be seen on any day. Every animal that used the river, except the elephant and the hippopotamus, remained alert to the prehistoric hunters whenever they approached the sand and mud flats at the foot of the banks. Indiscriminate hunters, crocodiles gave any haven the potential of becoming a trap.

In the thickets that lined the bank hornbills skitted in their heavy way between trees and bushes, and tiny kingfishers rested between diving attacks at the river's surface. Lilac-breasted rollers posed on the highest branches of the bushes they chose and flashed color whenever they took flight. Bird calls provided a constant chorus from just before dawn until just before dusk. In the higher, secret places of the tree communities, eagle owls peered out and waited. Several secretary birds nested in the area. Their interest was not the river but the flat savannah that stretched away on either side. It was there that their reptile prey could be taken by their stalk-and-stamp technique. Ground squirrels, mole rats, mongooses, and a whole array of smaller mammals provided the hunting birds with more than they needed. A huge python had several resting areas near the river. There was always enough for him as well.

The river was a rich place where food and water could be found in reliable variety and quantity even when the broad plains on either side began to dry out, even after they had turned to dust waiting for the rains. In every sense the river for its entire length was an oasis.

A single female leopard had a regular route along the western bank of the river, which she patrolled in one- to two-week cycles. There were always enough hoofed animals using the river, and more than enough heavy tree limbs that reached out across their trails made ambush easy and stalking rarely necessary. Males came regularly enough for her to mate when she wanted, and she had raised three litters of cubs within a hundred yards of the water itself. She was sleek and contented and extremely territorial. There were even days when she stalked the sandy banks and killed a young crocodile or two in what seemed to be spite. She carried the babies dangling from her jaws after the killing bite, but inevitably put them down and sprang up the bank toward her familiar trees without ever feeding on her victims. Perhaps, in some primitive way, she was thinning out the competition, reasserting her command of the river and its immediate surroundings.

When her movements coincided with the coming of the elephant she selected a high branch and lay there glowering down at the beasts she dared not approach in peace, much less stalk. Grumbling to herself, she waited until they were done and had gone back up their sloping trails before even moving from one branch to another. Under no circumstances would the elephants have tolerated her presence. Elephants have an aversion to all large predators, threat or no threat. She was a fairly large cat, just over 150 pounds, while the elephants using the river weighed up to 7 tons, between ninety and ninety-five times her size. There was no contest, none at all.

Across the river from the general area used by the leopard, a pride of lions centered much of its activity. Its territory, combining both the wet-season and the dry-season ranges, covered about sixty square miles, but when the savannah farther out failed it, when the great herds had moved south with the rain, the pride moved back toward the river and used what was to be found there until the herds would come again.

There were seven females and eleven cubs in the pride when Simba and Ol-Kurrukur began moving in their direction. The year before a large and aggressive male had ruled the pride, but he had fallen ill with a parasite infestation and the sicker he had gotten the less alert he had become. One evening, just at dusk, he had been down on the sand flats lapping water from one of the streams that marked the course of the parent river. He had died an unlikely death, held underwater in the jaws of a twelve-foot Nile crocodile. The younger cubs now with the pride had been fathered by an itinerant male that had passed through the pride's territory. Finding it unguarded, he had remained long enough to mate. The older cubs were the offspring of the now-dead ruler.

Before the aggressive male had driven him off a very old male had lived with the pride and, although he had mated often, few cubs had been produced. After being driven off he had vanished to the east and was never seen again. Huge packs of hyenas lived on the plains he had to cross. He never reached the far side.

Simba and Ol-Kurrukur became aware of the pride several days away from its position near the river. There were old kills that still carried unmistakable lion signs and scents. There were old scrapes, marked bushes, and fresher female and cub smells. The two wanderers were grown now, at full size and a year into their sexual maturity. On the second day of their approach they picked up the scent of a lioness in heat and increased the length and speed of their strides. Several times they broke into a trot, and, forgetting how lazy they really were, they aimed their course on a straight line toward the river, passing several easy hunting opportunities along the way. For the first time in their association, food was not their primary concern.

On the third day, toward evening, with the birds in full song as the cool winds began swirling down the river valley, Simba and Ol-Kurrukur came upon a new sign, scents laid down no more than an hour or two earlier. It was clear this time that they were right—no male was in attendance and none had been for a long time. Boundary marks had been absent or barely discernible during their trip in. There were none now, just the overpowering scent of a receptive female. For the first time since their meeting Simba and Ol-Kurrukur began jockeying for position. They did not look at each other, but each tried to rush ahead, to lead the way. Ol-Kurrukur was already grumbling and making deep chest and throat sounds. Simba followed suit, and two grunting males soon closed in on the pride.

They passed one female with two cubs, then a second with three. The males knew they were there, but that was not what they were after. In the distance a female stood, stretched, and yawned, staring at them with large golden eyes. When she had finished her stretching routine she continued to hold up her tail instead of dropping it as she normally would have done. It was an invitation and, as she turned her back on the advancing males, a presentation.

Simba was in the lead when suddenly, with no warning, his now longtime companion Ol-Kurrukur landed on his hindquarters and bit deeply into his flank. With a roar of pain and rage Simba spun, throwing Ol-Kurrukur off balance long enough to roll him onto his back. Ol-Kurrukur struck upward and back with his hind feet, claws bared. Simba sustained several deep gashes, but his teeth had already met in Ol-Kurrukur's shoulder. Both cats were roaring, their anger as real as it was monumental. In all things they had shared, but with a presenting female at stake the decision finally had to be made. Which male was to be dominant would be decided once and for all time by the outcome of this battle. Although the males would again be friends and constant companions after the victor had mated for several days, this moment, this situation and opportunity, made them blood enemies.

Ol-Kurrukur managed to rock loose and roll away from Simba, who now pursued him through the brush. A female and her cubs who lay in their path moved away. Other females were alert and watchful. If necessary they would move as much as a mile away until the tempest had passed. Simba caught up with Ol-Kurrukur and landed on his flanks just as the golden-maned cat had done to him barely moments before. Again they were down and rolling with each cat struggling to regain his feet and tower above the other. Retention of a superior position had a great deal to do with the outcome.

A number of birds sailed down from their perches, crossed the river cut, and sought more peaceful places to rest away from the squalling cats. Vervets on the far bank fussed about the disturbance among the neighboring lions. For no apparent reason several squabbles broke out among the monkeys as well. An eagle owl looked down and rolled his eyelids, seemingly in disgust. The noise from the lions increased in volume and so did the fussing among the vervets. Almost two miles away, the female leopard grumbled to herself, no longer able to ignore the lions, something she always tried to do.

The dominant position went first to one cat then to the other, but clearly Ol-Kurrukur was weakening faster. From somewhere deep inside, Simba had found the extra strength, the greater endurance that would make him master. Finally Ol-Kurrukur broke away and ran in the opposite direction from the still taunting female. He ran nearly half a mile, twice the distance Simba was willing to pursue him, before flopping down in a shaded area where he began to lick his torn and bruised hide. Several of the wounds were deep and would take weeks to heal. Infections would develop, but he would survive them all. His day, not his career, was lost. He had, in fact, inflicted as much damage as he had received.

Simba limped back from the pursuit and walked directly up to the female who was still lashing her tail high above her back, keeping her front quarters slightly lower than her hind. Simba walked into her, almost knocking her off her feet. He bit her gently on the back of the neck, placed a lame left leg across her back, slid back down to a position beside her, and pressed his flanks against hers. Together they walked straight ahead, leaning in on each other until the bushes closed behind them. Nearly a mile away they emerged from the bushy area, Simba in the lead, the female following behind. Another mile passed before they reached a second bushy area and then Simba turned and butted his head against hers. She returned the butt, rubbed her cheek against his, and then snarled and struck out at him with her great right front foot. He dodged the blow, returned it half-heartedly, and bit her on the neck again. In future encounters with females in heat Simba would mate in the center of the pride's resting area, but the battle with Ol-Kurrukur had left him tense and he could not be sure an attack would not come at any moment.

Ol-Kurrukur did not appear as that night and throughout the next day Simba mounted the female over thirty times. Simba's animation at the first sign of an estrous female, the hurry to reach her, the terrible battle he had fought with his friend, and all of the courting and mating activity had exhausted him, and he finally slept deep and long. Toward dawn after the second night he began again, mating fourteen times before noon. On the third day, after mating several more times but with ever-diminishing enthusiasm on both their parts, they turned toward the place near the river where they had left the pride. Within forty minutes they had rejoined the other cats that now included Ol-Kurrukur. Simba walked up to him, the two males butting heads as they had always done, and then headed toward some bushes a dozen feet away. He flopped down onto his belly and rolled over onto his side.

Ol-Kurrukur stood and approached the female that had become Simba's temporary mate. He sniffed and lay down next to her. He was still badly crippled by several wounds that had not yet closed. Still, a few hours later, he did mate with her several times even though her estrous had passed—there were no eggs left to descend and meet his sperm.

Although both males would on occasion share the fatherhood of a litter, this one, the first that would be born after their arrival, would belong to Simba. He would take little interest in the cubs and no special interest in the female until she came into heat again, but black-maned males would be born to this pride. Although all of them would be driven off just as Simba and Ol-Kurrukur had been, they would carry the strength and perfection of their father with them. It would be no less true of the cats with manes of gold. They, too, would carry the power of perfection with them and bring it to other prides. Simba and Ol-Kurrukur would remain in place as dominant males and would seldom have to hunt again.

❧ 26

Man's Ability to Care

From
A Celebration of Dogs

And for the two last selections on place in the lives of animals I will take license. In fact, I will play with words here for my last species under consideration is the domestic dog. Its place? In our hearts. There is a perfectly valid scientific explanation why this is so, why we have taken our companion animals, basically the dog and the cat, and made them the center of their own little universe. But much of that can be found in the selections that follow from A Celebration of Dogs. *(At the time I'm writing this,* A Celebration of Cats *is at the printer, carrying the stamp of Simon & Schuster.)*

So, in our hearts, naturalists and nonnaturalists alike, millions upon scores of millions of us, there is a place that two species have carved out for themselves. (In these selections the dog will stand for both.) Our relationship to them is of vast historical and economical significance. Emotionally and intellectually that relationship is a joy to behold, truly something worthy of celebration.

Before there was a man-dog relationship there obviously had to be both men and women, and dogs. Men and women came first, at least before the dog we keep today as a companion animal. The far older dogs of Asia and Africa, as I will explain, probably played no role at all in the coming of Spot or Fido.

The epoch we call the Eocene began approximately fifty million years ago and lasted for perhaps twenty million years. It was during that time that a beast known as *Miacis* emerged. It was a combination of weasel, cat, bear, and dog; in short, it was an extremely primitive carnivorous type. It had a future, however. It had promise because it contained a lot of good mammalian ideas.

Starting thirty-five million years ago, for a period lasting almost fifteen million years, there was the epoch called the Oligocene. It was during that period that *Miacis* gave way to several even better ideas. There was *Daphnaenus*, an animal much heavier than its ancestor *Miacis* had been. From *Daphnaenus* would one day come all of the bears. There was a lighter-boned descendant of *Miacis* known as *Cynodictis*, and from it would come the dog family. It was at that point, when *Daphnaenus* and *Cynodictis* coexisted, that the dogs and bears split for all time. *Cynodesmus* was next in line after *Cynodictis*. It was more doglike yet. The weasels and cats that had also arisen from *Miacis* were headed off on their own, like the bears. The hyenas, more catlike than doglike anatomically, were also split off and were edging away from the line of canine descent. The meat-eaters were becoming more and more differentiated, more specialized. The dinosaurs were gone, the mammals had inherited the earth. With their superior brains they were exploring every niche and trying on all shapes and sizes.

The Miocene epoch dawned about twenty million years ago and was destined to last for close to thirteen million years. Man was still about seventeen million years in the future, but *Cynodesmus* had already evolved to *Tomarctus*, and from that animal the wolves, foxes, wild dogs, and jackals would eventually arise. In at least one of them our pet dog of today was hidden. All kinds of canines came and went, new ideas replaced old ones, and the line continued its refinement. Parallel tracks were laid down, but it was within the wolf that lay the dog man would one day extract, polish, and eventually cherish. That is where the dog's story starts, roughly fifty million years after the mists of eternity produced *Miacis* and a couple of million years after the man-idea pushed apes upright, moved their spinal column forward under their skull, straightened out their bowed knees, and made their thumbs grow. We know rather less about the details than we would like, but we have the results all around us, and that, surely, is far more important.

To pinpoint the actual beginning of the man-dog relationship requires a good bit of conjecture, as do most things that happened between one hundred and two hundred centuries ago. In most parts of the world the dog was probably the first domestic animal man extracted from the living forms around him. There seems to be little argument on that point. Most likely it happened somewhere in the Middle East or southwestern Asia. The date suggested for the event is often fourteen thousand years B.P. (before the present). That may be a modest estimate, since man had purebred dogs seven thousand years ago in America, and the remaining seven thousand years seems a fairly short time for that level of technical accomplishment to be arrived at in the Stone Age and that much traveling to have been done. In fact, how people who couldn't write and probably had more fleas than their dogs mastered selective breeding is thoroughly confounding.

The first issue, however, is how and why it happened. Man the carnivore—more properly the omnivore, because he was surely hunting, gathering, and eating everything in sight, including his neighbor in some instances—appears to have suddenly reached out and taken a fellow carnivore and therefore competitor into the cave and made him a partner. All this with no history or tradition of domestication. It almost seems as if there had to have been some kind of revelation.

There have been any number of romanticized views of how it came to pass. One day Mrs. Flintstone supposedly said something to the effect that the kids would love a puppy and Fred should bring one home from work. Unlikely. Slightly more plausible, but still probably less than likely, is the idea that some small species of wolf not unlike one still to be found from India to Israel and probably once even more widely distributed than that (*Canis lupus pallipes*) began hanging around human dwellings to get the offal tossed aside by the hunting families and clans. Eventually, the theory goes, it became a habit for man and wolf to stay near each other, and it stuck and went on from there. Possible, but again I think unlikely.

What does seem likely is that man had a hard time making ends meet and brought home everything he could chew and swallow. There was a constant plaint back at the nest: more food, more food! A raid on a wolf den when the adults were away or had been neutralized with a shower of rocks would have periodically yielded a batch of tender puppies. Dogs are still eaten in Asia, so I have no problem with this speculation so far. Even cavemen would have discovered that meat goes high if it is dead and cannot be kept cold. Meat on the hoof stays fresh. It is my guess that pre-dog man began keeping puppies from kidnapped wolf litters around for food as needed. It is not difficult to imagine that sooner or later some kid would ask some father to spare a puppy. It was probably done as a temporary measure, but a habit came into being with which we are still enchanted. That little scene in the cave, however, is still a long way from purebred dogs genetically engineered to fulfill a job like sheepherding or game tracking or guard duty. There was still a way to go. A wolf puppy cannot grow up to be a dog no matter how much you love it.

And what does the record show? For a long time it was assumed that there were two ancestors to the dog, the wolf and the jackal. That has pretty much gone by the way, although wolves, jackals, and domestic dogs can all interbreed. There is a lot of liberal democracy in the canine genetic package. But the wolf generally gets full credit today for giving us our dogs. Large wolves come from the northern latitudes and smaller wolves from southern ones, so the southern wolf, as suggested a moment ago, again generally gets most of the credit. I personally believe that the larger northern wolf figures in the ancestry of northern spitz-like dogs, but for the mainstream of domestic dog development I would accept that little southerly wolf, because there isn't really very much more we have to go with.

Where? Certainly not in any one place. An idea that has been good enough to last for one to two hundred centuries was almost certainly good enough to have more than one inventor. There were, in all likelihood, many places where the socialization of wolf began, and many places where the idea caught on.

When scientists dig up ancient human sites, how do they know if they are finding wolf remains or dog remains? That certainly is a significant point if one wants to claim that a particular race of people had dogs. It isn't as difficult as it may seem. Strange things happen to canines when they are domesticated. Sad to relate, but unavoidably true, is the fact that their brains shrink. A dog's brain is 20 to 30 percent smaller than that of a wolf the same size. Teeth become more crowded as muzzles shorten. Wolves and dogs, even in the earliest stages of domestication, can almost always be distinguished from each other. Paleontologists have little trouble making a determination.

Unexpectedly—bewilderingly, in fact—the single earliest domestic dog so far known was found in the New World, not the Old. In Birch Creek Valley in Idaho there is an archaeological site known as Jaguar Cave, and there, roughly 10,500 years ago, there were men with small dogs that had short, broad muzzles. The premolars of those dogs were crowded, proving to virtually everyone's satisfaction that this was indeed a dog and not a wolf. At least as interesting is the fact that these dogs show every sign of having been descended from the Old World wolves, not any wolf species found in North America. They were a long way from home. These were dogs that came to this hemisphere from Asia, although it was probably much farther west, where Europe meets Asia, that dogs first began to emerge, those small-brained wolves whose teeth were being crowded into an ever shortening muzzle. There are no details about that incredible trek. It is a long way from Jericho to Idaho. What happened to the dog along the way, and what happened to man?

In Europe itself there are clear records of dogs in Germany, in a place called Senckenberg Bog in Frankfurt am Main. They date from about 9,500 years ago. In the British Isles there were dogs in a place called Starr Carr. The dogs in England seem to have been more advanced than their German counterparts, although the German and English finds are from about the same time. Since it is highly likely that early dog owners allowed the influx of wolf blood from time to time (Eskimos have done that with their sled dogs to the present day), the relative primitiveness of a specimen or two from a hundred centuries ago may not tell us very much about when dogs started or where. We are dealing, it must be remembered, with a skimpy record and a lot of guesswork.

In Anatolia, Turkey's hind foot in Asia, there were dogs at least nine thousand years ago, particularly at a place called Cayonu. It is possible, just possible, that in some isolated areas sheep or goats or perhaps even both were domesticated a little earlier than dogs. Generally, though, dogs were the first animals domesticated, or rather wolves were socialized and dogs later evolved from them.

What those early dogs looked like is hard to say. The domestic canine is an amazingly flexible genetic package, and all manner of variations have come and gone. Those at the beginning are particularly difficult to identify. In the newer Stone Age sites in Switzerland there appears a dog variously known as the turbary dog or turbary spitz. It was spitz-like; it was quite small, certainly smaller than the wolf from which it was descended; and it had a spacious brain case, which probably meant it was a smart little animal not far removed from its genetic origins. Its remains are found in Neolithic sites scattered all over Switzerland and in other parts of Europe as well. There were at least two other larger dogs in Switzerland at the same time, indicating that even Stone Age man had his choice of breeds. There were a variety of dogs in Russia at the same time, perhaps already doing different kinds of jobs for their owners.

Thus, ten thousand years ago men in several states of technological development had dogs not of one kind but of several. Some scientists believe they can ascribe work types to these dogs. Some seem to have been bred for hunting and others for herding. Some of the individual conclusions drawn may be less than totally sound, but the story as a whole seems clear. A hundred centuries ago, men of very little technological sophistication and probably next to no knowledge of biology were selectively breeding dogs. How is a mystery; why seems obvious. If we knew more about the history of dogs, it seems clear, we would know more about the history of man.

It is said that mankind now doubles his technological skill every five years in a kind of intellectual inflation. At the time I am speaking of, one hundred centuries ago, it supposedly took man thousands of years to achieve the equivalent doubling. Allowing that such figures are probably fairly loose, it would still mean that if ten thousand years ago man was selectively breeding for breed types he must have had dogs around for a very long time before that. When his canid companions actually crossed over and became dogs can never be established with precision, so a good part of that earlier time may have been a matter of keeping cleverly socialized wolves. My conclusion is that man and canine have been companions for a period that could cover two hundred and fifty centuries, twenty-five thousand years. Very few human cultural complexes would have preceded that. Dog-keeping or at least wolf-about-to-be-dog-keeping may be the fourth oldest of all our cultural activities. Hunting and gathering, lumped together as a single activity, and perhaps storytelling came earlier. Fire almost certainly came next. Man had to eat before he could keep pets. As suggested earlier, it was probably his eating habits that triggered the first pet-owning that man ever attempted.

.   .   .

It would be absurd to think that man's interest in dogs remained strictly culinary or in any other way utilitarian for any appreciable length of time. Archaeological sites yield toy breeds and companions very early in the dog's career, and all of the rest of the history of man and dog over much of the world reveals a phenomenon known as bonding. Men bond with men, with women, women bond with men, with women; that is obviously known. What only animal lovers themselves may really appreciate is that human beings bond with nonhuman beings. It must be acknowledged here that what I am about to say for dogs is as true for cat lovers, horse lovers, anyone who finds a nonhuman companion fulfilling. But dogs are pack animals (just as horses are herd animals) and have a compelling instinct to fit into a structured situation. That and their size make them easy for us to relate to.

Amazingly enough, this fact was barely examined until recent times. The Old Testament did suggest, "A just man regardeth his beast," and Jews and Christians ever since have been trying desperately to show that this proves a humane ethic comes down to us from the Bible while all the while wallowing in a sea of animal agony caused by or at least allowed by man. In fact, until recently, dogs were taken for granted. They were always here; no human being alive could ever remember anyone who could remember a time when they weren't here, so in Western cultures, at least, next to no thought was given the matter. Either you liked dogs or you didn't. Either you had dogs or you didn't. There was nothing more to it than that unless you bet on them in fights or races, neither event inspiring sentimentality. But stop and think about that for a moment.

As suggested earlier, a vast array of cultural complexes arose and most died since the first dogs were selectively bred during Mesolithic times. Human sacrifice was a perfectly valid form of propitiating the gods, or at least the natural forces that seemed to have the power to make things good or bad in a most whimsical fashion. It arose after dog-breeding had been established, because there wasn't any religion as we understand it before that, but that has gone by the way. Slavery came and went, although that was a valid economic device as long as you were the slaver and not the slavee. Cannibalism has vanished from a protein-starved world, but dog-keeping has hung on. Only one major religion, Islam, has not been accommodating to dogs. Virtually no forms of government have found dogs a poor idea except in isolated instances and then usually for a short period of time. Iceland once did not permit pet dogs; you had to have sheep and need dogs as helpers in the pasture. In short, dogs have been acknowledged tacitly or by decree as a good idea almost as far back as any shovel can reveal.

What is behind it? Bonding. Call it love, call it whatever you wish, but don't suggest misanthropy. Psychiatrists have so long since proved that dog lovers are not people haters that whoever first made the silly suggestion must be bruised from turning in his or her grave. And what is the nature of the bond?

First, dogs are nonjudgmental. It doesn't matter where you are in your own personal development, nor has it mattered where you have been culturally; dogs simply don't pass judgment on you the way all the rest of life and all of your other companions seem to. Miss a spear throw, the gazelle gets away, going home empty-handed? Your dog will feel the pinch in his tummy as much as anyone else in the cave, but the dog won't whisper behind your back and blame you. Reputation and respect were probably as important in a cave as they are in a condominium. When men and women fail at something they generally have a reserve of paranoia to draw upon. We are all capable of being paranoid, and it never feels good. Dogs do not summon up that reserve. Flunk a French exam, hell coming at you from all quarters? Your dog couldn't care less. Get fired, cheat at cards or on a higher plane? Your dog will give you exactly the same greeting. The nonjudgmental quality of animal companions, the dog always having been the most popular of that group, is a major source of strength for the glue that binds us to so unlikely a friend. Unlikely? Four-legged, dependent, capable of transmitting one of history's most dread diseases, rabies, nonverbal, of a much lower intelligence, yet adored. Unlikely, but adored to the point of near ecstasy by some people and some cultures.

The fact is that people do better when bonded. People who live alone do not usually live as long as people who have a satisfactory relationship with someone or something else. Single people die earlier than married people, on the national average. Single people who live alone get sick more often than solidly bonded people and suffer more severe cases of the diseases they get. All of this has been known for some time. What no one seemed to notice before a couple of years ago was that people bonding to dogs (and other pets) experience pretty much the same effects. There may be differences in degree, but dogs do very well by us.

In 1980, four authors at the University of Pennsylvania published a paper called "Pet Ownership and Survival After Coronary Heart Disease." The authors, Dr. Erika Friedmann, Aaron Katcher, James Lynch, and Sue Ann Thomas, had come up with some startling facts about heart-attack victims.

The findings showed that if you had heart trouble you would probably have a much better chance of surviving for a longer period of time if you were bonded to a companion animal than if you weren't. Fifty-three people, each of whom had had a single heart attack, were placed in one group. Each of them either had a pet or was given one. Of those fifty-three people, three died within the first year of the study from a second heart attack. In the control group, thirty-nine people without pets were watched. In that same one-year period, eleven died of a second heart attack. Three out of fifty-three as compared with eleven out of thirty-nine: those are not mortality figures easily ignored. It could have been argued that the people with pets survived because they walked their dogs, and walking is very good for most heart-attack victims. To offset that possibility a number of the fifty-three were given pets that made no physical demands, like cage birds and guinea pigs. What was shown was that people are much healthier and much happier when involved with animals than they are living alone.

What I find amazing is that anyone finds this amazing. As a child of ten I am certain that if I had been told of another kid down the street who was in a wheelchair and who was isolated and lonely I would have automatically said, "Get him a puppy." Of course, I was always up to my ears in puppies, kittens, turtles, snakes, and about everything else that could be crammed into a house and yard.

Studies in Maryland in the past few years revealed that people with explosively high blood pressure had their pressure drop almost immediately if a cat or dog was brought to their side or placed in their lap. People do reflexively pat companion animals when they come into contact with them, and patting seems to have an immensely calming effect on people suffering from hypertension. I don't believe all the science involved in this effect has been worked out. Whether the tactile sensation of fur on a warm body does it, whether the attention of another living creature does it, just isn't known, but there is more than enough evidence to suggest that Mesolithic man was on to something big, and even if statistics to support his actions and the science to explain it were ten to twenty thousand years in the future, he hung in there. How much of man's interest in the dog was based on some innate common sense cannot be told. More than utility was involved, much more. The bone record supports that. Of course, in the case of those early toys, affection could be considered a form of utility. We are just beginning to really understand another element in the man-dog bond, the psychological aspects, those that are strictly emotional. Studies have shown that people can be helped emotionally by dogs to an enormous degree. There have been some very graphic examples in my life and within my view to bear out the critical nature this bond can assume. I recognize it in myself and family, but I have seen it in a very vivid way beyond my immediate intimate circle.

When I was fourteen years old, and large for my age, I got a job at the famed Angell Memorial Animal Hospital. Located on Longwood Avenue in the heart of Boston's vast, sprawling medical district—Beth Israel, Boston Lying-in, Boston Children's, Harvard Medical School, Massachusetts Dental, and a host of other institutions were all within a few blocks of Angell—it has been consistently rated one of the finest veterinary hospitals in the world. It was operated then in its old setting and is today in its new home on South Huntington Avenue by the Massachusetts Society for the Prevention of Cruelty to Animals. The MSPCA is one of the oldest humane societies in the Western Hemisphere.

The first assignment given to the eager and somewhat brash new boy was, of course, cage cleaning. After my initial trial period I graduated to the waist-high bathtubs in the grooming section, watching my hands and forearms turn into prunes as I worked. I bathed, clipped, and de-burred dogs all day long. I seriously questioned if my back would ever be right again. A slight forward tilt for eight hours a day can do one heck of a job, even on a teenager's sacroiliac.

Promotions came in proper order, and after passing through some less rewarding departments (like the euthanasia room) I ended up in a white jacket in the clinic helping the staff veterinarians, or at least doing menial chores that made their lives a little easier. I hoisted animals on and off examining tables (after first lathering the stainless-steel surfaces with antiseptic solutions), held animals while they were examined and treated, guided clients in and out, and took dogs and cats back and logged them in if they had to be hospitalized. It was a rich experience for a boy just turning fifteen. Heading for work after school and on weekends didn't seem a chore, really. I was beginning to focus on a career. At that point I was determined to become a veterinarian.

One day the veterinarian on duty groaned as he pulled a card from a wooden holder on the wall. It was a familiar card, one side completely filled in with very little room left on the reverse.

"O.K., ask Mr. Jones to come in." (Jones was not his name, but that hardly matters.)

The client was a tired man with gray skin. He gave the impression that life had not been an unmixed blessing. His weariness hung on him like an ill-fitting shawl. In his arms, wrapped in an old but obviously frequently laundered blanket, was a positively ancient Boston terrier. The old man put his old dog on the table gently, obviously with love, and looked up at the tall young veterinarian, who towered over him. There was some hope in the old man's eyes, but not really very much.

The veterinarian had seen this dog before, often, but he went through the gestures of examining him: stethoscope to heart, palpation of the abdomen, a look in the mouth and down the throat, a quick glance in the ears.

"The new medicine doesn't seem to be helping very much," the old man fairly croaked. "It's hard to tell, but there doesn't seem to be much change."

The veterinarian looked down at the floor for a moment, then put his foot on the rung of the examining table. He took a deep breath and leaned forward, prepared to make the speech he had made so many times before.

"You know there is never going to be any change,

Mr. Jones. You know very well I am giving you medication for . . . " The veterinarian glanced over at the record card. ". . . for Gutsy to humor you. I can't be any blunter than that. Your dog is eighteen years old. He is blind, deaf, incontinent, he can't walk, he is frightened and in pain, and you are not being nice to him by keeping him alive this long. In nature he would have died long ago."

The young doctor was really being as kind as he knew how to be, but he had been begging Mr. Jones to have his dog put to sleep for over six months. Mr. Jones, however, was back every week trying new medication, always with the faintest hint of hope in his eyes. We all knew him. We all understood his plight, or at least we thought we did. There wasn't a veterinarian or a kennel helper at Angell who hadn't been through the same scene many times before.

"You don't think there is anything you can do?" The same question was asked every week.

"No, I *know* there is nothing we can do. We can't turn back the clock, and time has run out on Gutsy. You are not being kind to him by prolonging this very bad period for him."

The old man thought for a moment, then shook his head. He had come to a decision.

"Well, if it must be it must be. I'll take him home and do it now."

The veterinarian reached out and took the old man's elbow as he bent forward to wrap his old dog up and carry him away.

"Don't do that, please. It is a very difficult thing to do at home. You won't have the right materials, and you will be cruel to him even though you are trying to be kind. Let me do it here. I have a drug that will work very quickly. You can stay and help me if you want. You hold him and I will give him an injection. He won't even feel it and it will be over instantly. You will be able to see that for yourself. Please, don't try it yourself."

The old man thought again for a moment, then looked up into the veterinarian's eyes. Having finally decided to put his dog to death, the old man was finding new strengths, strengths that obviously had been eluding him for a long time.

"My wife died almost twenty years ago. I never wanted to remarry. We had just one son, and he married a girl in China. They have three children, but I have never been able to go there and they can't afford to come here. We're half a world apart. I don't know why God worked it out this way for me, but Gutsy is all the family I have had for a long, long time. If he has to be killed, I'll do it. It is up to me. Thank you, doctor."

The veterinarian started to protest again, but the old man had gathered his dog up in his arms and was at the door. I was in the process of opening it for him when he turned back.

"Gutsy and I really do appreciate everything you have tried to do. We understand, truly we do."

The old man was gone, and the doctor was shaking his head as he reached for the next card in the wall rack.

We read about it the next day. It wasn't front-page news, by any means, but it did make most of the Boston dailies. The old man had gone home, stuffed paper under the door, sealed the windows, and placed his rocking chair in front of the stove. He turned the oven on, but he didn't light it. I am sure he was rocking slowly, perhaps humming reassuringly to Gutsy, as they both went to sleep. I cried. I think the veterinarian did, too. I know some of the other kennel kids did, and I resented those who didn't.

The old man who committed suicide with his dog was not neurotic. He was an even-tempered, gentle, reasonable man whose one failing seemed to be his inability to let go of hope. Even when it was all stacked up against Gutsy he dared hope. Judging from the way he dealt with the veterinarians I saw him with, and with the kids who worked in the clinic, he was hardly a misanthrope. He was a nice old guy who had discovered, to his joy, the role a dog can play in a human life. His suicide was sad but understandable. He was not a "nut," he was not antisocial. He simply could not live with the inevitability of the truism that every pet taken into our lives is a tragedy waiting to happen.

Dr. Boris Levinson has been a pioneer in exploring the role of dogs in the lives of children. He has published two books on the subject, *Pet Oriented Child Psychotherapy* and *Pets and Human Development*, both published by Saunders. These professional publications are by no means the only studies done, for in this country as well as abroad the subject has been coming more and more under scrutiny.

The value of the bond between child and pet is based on a number of factors. There is again and perhaps in many cases foremost the matter of the pet's being nonjudgmental. Children are always under the gun to learn something, do something, become something, and they need constant approval. They seldom get enough. Kids are not always certain they are going to be able to live up to everyone's expectations of them. Many are frightened by what lies ahead, and why shouldn't they be? The most reliable source of approval for most youngsters is apparently a pet, and dogs are generally the most demonstrative of these. Kids can rely on their dogs to always pay attention to them, always approve of them, always find the time and energy to interact with them.

From the societal point of view, kids and dogs mix in another highly beneficial way. Kids are naturally egocentric and irresponsible. They react positively to whatever feels good and negatively to whatever does not. It takes some doing to bridge out of that hopeless mode and become a reliable social interactor. Dogs can help tremendously, because they can help responsibility and selflessness seem much less awesome. For many children, their first responsible actions, the first things they do on their own that are not for their own gratification, are in helping to care for a pet. Wise parents take advantage of this and encourage even tiny kids to check the water dish, and, in time, walk the dog. Certainly, kids can grow up to be responsible adults without having pets. It is just a whole lot easier, very often, if they do have them.

In Brewster, New York, there is a school for emotionally disturbed children called Green Chimneys. Approximately 40 percent of the students' time is spent on the school farm helping care for animals. Kids who had never been exposed to animals suddenly find themselves members of 4-H, find themselves going to the animals when they feel their own pressures building up inside. For many of them the farm at Green Chimneys is the first device other than drugs that has ever worked in getting themselves under control, keeping them from blowing up like volcanoes. Although dogs are not the focus at Green Chimneys, the point is still made. Children and animals are an incredibly healthy mix.

In Bridgeport, Connecticut, there is a home for alcoholics. I visited the institution and found that one of the keys to its program of rehabilitation was the dog, specifically the Labrador retriever. One patient-inmate of twenty-five sat cross-legged on his bed patting his Lab. He sobbed as he confessed to me he had been a skid-row bum since the age of sixteen. "This," he said, nodding toward his dog, "is the first responsibility I have ever accepted in my entire life without panicking." At a school called Lincoln Hall in northern Westchester County, New York, the brothers who ran the school for PINs (People in Need of Supervision; we used to call them juvenile delinquents) assigned different breeds of dogs to each of the bungalows. The twenty-four students and the dogs got along just fine, and the brothers felt the dogs helped enormously in rehabilitating the troubled youngsters.

The Lima State Institution in Lima, Ohio, houses men who have been classified as criminally insane. That is a kind of ultimate category. When you can't make it in a regular insane asylum, when even the penitentiaries don't want you, they send you to a place like Lima. The men arriving there have an attempted suicide rate of about 85 percent. Not only have many of the men tried to kill themselves one or more times, but many of them are self-mutilators. One of the men I interviewed had been kicked out of his former residence, the state penitentiary, because he had swallowed razor blades. There are six sections or wards in Lima where the men are allowed to keep pets. In seven years in those six sections there has not been one suicide attempt. This suggests that animals have a profound effect on mental health or at least on self-control. Where Valium failed in the other wards, pets succeeded in those assigned to them.

A specific event at Lima reveals how much animal companions can have to do with social control. One of the inmates went berserk in a recreation room. As so often happens with a group of people who are individually marginal in their social mechanisms, the insanity quickly spread. Men began beating each other up for no apparent reason, and then the room itself became the focus of their uncontrollable wrath. The television set was demolished. The curtains came down, the wood and glass windows disintegrated; only the steel bars beyond kept the explosion contained. Furniture sailed through the air; even a billiard table went down in splinters, although how that was accomplished without a bulldozer remains as much a mystery as the cause of the mayhem itself. When the orderlies and guards came pouring through the door and had wrestled the last inmate to the floor, only one thing remained standing amid a sea of rubble. Only the most fragile thing in the room was untouched—a fish tank.

The reaction of a group of frenzied criminal minds to a small colony of fragile tropical fish may not seem like a major reflection on man and dog, but it is. It shows again what happens when human beings and animals bond as the keepers and the kept, and what is involved in the companion animal.

There was a remarkable case in Holland related to me by one of the principals involved. He is an industrialist with factories and businesses all over the world. One of his enterprises is located in the extreme northern part of the Netherlands, sticking out into a cold, gray sea. The weather there is almost always dreadful, and people stay inside much of the time.

The man's business there is a large factory prefabricating concrete buildings, often entire settlements. He was awarded a contract for a U.S. Air Force base in Greenland. It was totally prefabricated at his plant, everything from barracks to chapel to mess hall to hangars and administration buildings. The base was shipped by sea.

His Dutch workers are models of skilled and semi-skilled labor. He has never had labor trouble, and the enterprise has turned a substantial profit. He decided to declare a bonus and called a workers' committee together to decide how the bonus should be distributed. He offered them a cash bonus, an improved insurance plan, a retirement plan, what did they want? The committee said they would consult with the men and report back.

Keep in mind that this was in a very desolate, remote area. The men rode to work on bicycles carrying their lunches with them. There was a long, gray, dull prefabricated dining room where they gathered to get hot tea, chocolate, or coffee and eat their lunches. It was where workers' meetings were held, where all breaks were taken.

The committee reported back. The workers had agreed—it had been unanimous—that they wanted the money spent converting one long wall of their dining room into a floor-to-ceiling aviary. They wanted green growing things and birds. They wanted to be able to share their free time with living creatures, things that would add color and life to their gray world. They wanted to see birds from foreign lands, things they could otherwise see only in books. They wanted to be able not only to see and hear those feathered creatures, they wanted to be able to care for them. And that is how their bonus money was spent. My friend the industrialist was amazed but touched. He doesn't spend time at that factory any more than he does at any of his others, but he is somehow pleased that that beauty was asked for and delivered as promised.

Again, birds on a peninsula in the North Sea are not dogs, but the issue, the premise, is the same. Something good happens inside people when they are able to cross a bridge out of themselves into a more natural, less neurotic, and far less demanding world beyond. Dogs remain the best or at least the more popular of all the available bridges outbound.

Kate has been confined to a wheelchair all of her life. The daughter of a powerful, even awe-inspiring, and very beautiful author mother, Kate has had a great deal to contend with besides her own physical inadequacies. Fortunately, her mind is extremely acute and she is naturally sweet and generous. Kate needed something of her own. The parade of noted visitors her mother's fame brought through their handsome Victorian mansion was not enough, and neither were Kate's intense religious convictions. Not at all surprisingly, the answer was the old faithful companion. From her wheelchair Kate was able to obedience-train dogs and show them. Ribbons and trophies and framed photographs testify to more than twenty years of involvement, and dogs still grace the home and enrich the life of a woman who from the beginning seemed very, very short on luck. No one can even imagine how many hours those dogs filled and how desperately they were needed.

It is interesting and revealing that although fate decreed that Kate depend heavily on her dogs for companionship and fulfillment, she remains perhaps the least misanthropic human being I have ever known. No one is filled with more love or is more forgiving of her fellow humans' faults than Kate. The keeshonds, Newfoundlands, whippets, and German shepherds that have filled so many niches in Kate's life have in no way deprived her of her humanity. Except among the already desperately neurotic, dogs have the incredible capacity of remaining in their own emotional place. We need them, we may depend on them, but they never really replace people. It has been my experience that the opposite is true. Healthy dog lovers tend to be better people lovers than any dog haters I have ever known.

Dr. Samuel Corson and his wife and associate, Elizabeth O'Leary Corson, of Ohio State University's College of Medicine, recently published a paper called "Pet Animals as Socializing Catalysts in Geriatrics: An Experiment in Nonverbal Communication Therapy." A premise is established early in the paper: "In industrial societies with high population mobility, the emotional trauma of economic and social isolation in the aged is often superimposed on an earlier layer of psychological stress associated with the 'empty nest syndrome.'" Further, the authors state: "The loss of economic and social roles induced by obligatory age-related retirement thus may accentuate the psychological stress of the empty nest syndrome and lead to loss of self-esteem and of the ability or willingness to maintain some semblance of independent functioning, socialization, and goal-directed activities." In a word, by life today old people are made to feel unwanted, useless, and therefore awfully lonely.

Having established their premise, the authors then go on to quote "Song of Myself" by Walt Whitman:

I think I could turn and live with animals, they
are so placid and self-contained,
I stand and look at them long and long.

They do not sweat and whine about their condition,
They do not lie awake in the dark and weep for their sins,

They do not make me sick discussing their duty to God,
Not one is dissatisfied, not one is demented with
the mania of owning things.
Not one kneels to another, nor to his kind that
lived thousands of years ago,
Not one is respectable or unhappy over the whole earth.

So they show their relations to me and I accept them,
They bring me tokens of myself, they evince them
plainly in their possession.

Armed with a premise and a rationale, they seek evidence of the value of dogs and other animals, although they do feel dogs make the best therapists. One aspect derives from the nonverbal signals health-care workers, doctors, nurses, attendants, and even friends and family send to the sick, the infirm, the disabled, and the mentally impaired. However hard we try not to do it we convey pity or disgust or concern or revulsion; in some way, generally through facial expressions and perhaps physical avoidance, we let less than perfect people know we find them if not repugnant then at least, well, less than perfect. That is apparently something that cannot be helped, because we usually don't know we are doing it. Dogs don't do that, and that fact can make the difference between sanity and near desperation in someone already beset by troubles.

The flow of negative signals apparently intensifies in the custodial setting, where the staff is inevitably overworked and the emotional needs of patients or residents overlooked. One way of breaking the cycle of negative signals that gives rise to increasing helplessness in a self-reinforcing circle is the introduction of attendants who can't participate—dogs, and to some extent other animals as well. Dogs are especially good because they are the most immediately and apparently responsive. (Not to be denied, however, is how well some cats can play the role.)

In addition to the nonjudgmental quality of dogs, there is another quality that works well for the elderly. Dogs maintain a kind of *perpetual infantile innocent dependence*. According to the Corsons, that stimulates in us a natural tendency to offer support and protection. It gives people a purpose, makes them not only feel wanted but *know* that they are wanted. It is not a game; no one has to pretend or patronize. Dogs do need human support and show gratitude when they get it.

All that brings us to another quality in the dog that makes it so endearing. Although the dog plays a role, it is not a conscious role player. A dog is utterly sincere. It cannot pretend, it cannot act in any way patronizing no matter how physically or emotionally incapacitated the human participant is. For a dog every human being is exactly the same, except some are more responsive and likable, from the dog's point of view, than others.

The fact that people do not have to be suspicious of a dog's reactions to them is in itself an enormous measure of potential mental health. You may question almost all or perhaps all of the people in your life, but you don't question your dog. People use you and pretend they don't, while dogs use you in complete honesty because they have no choice, and they have not an ounce of deceit in their soul nor self-consciousness about any of this. It is certainly true that dogs have been symbols of evil and in mythology have played a wide variety of roles, but on a one-to-one basis they are not suspect, and old people, frightened people, alienated people, and people alone for whatever reason are frequently suspicious. To be able to relate without having to doubt, to be able to love or even just like a whole lot without having to fear rejection, is a source of comfort more fortunate people cannot begin to imagine. We can't, after all, imagine a toothache, not really; and we certainly can't imagine the terrors of lurking paranoia. A dog may be a nice pal to you, but it may be utter salvation for someone less fortunate.

Few if any of these factors could have been known to the people who first socialized *Canis lupis pallipes*, but then little was known to them about vitamins, proteins, and carbohydrates, yet they knew how to eat. Even people who do not understand the biological connection between sexual intercourse and pregnancy still engage in sexual intercourse, because nature, anticipating the essential nature of the act if the species was to survive, made the act pleasurable and therefore automatically repeatable whatever the biological sophistication of the participants. People have sex because it feels good, they eat because it feels good, and for a very long time they have kept companion animals because it feels good. The intellectualization of all three came long after the invention.

Without doubt the ways in which human beings bond to nonhuman creatures will be under investigation for decades. Ph.D.s will be awarded to the investigators by the score, and that is all to the good. The status of the dog and of all companion animals can only be raised by such studies, and that will come at a time when dogs and other animals, wild and domestic, can use all the help they can get. Without doubt, adults and children will benefit, because controlled programs using animals can have remarkably beneficial effects. In the complex world of the man-dog interface, things are looking up.